ENERGY SCIENCE, ENGINEERING AND TECHNOLOGY

SECURING GUAM'S ENERGY FUTURE

ASSESSMENTS, STRATEGIES AND ACTION PLANS

ENERGY SCIENCE, ENGINEERING AND TECHNOLOGY

Additional books in this series can be found on Nova's website under the Series tab.

Additional E-books in this series can be found on Nova's website under the E-book tab.

ENERGY POLICIES, POLITICS AND PRICES

Additional books in this series can be found on Nova's website under the Series tab.

Additional E-books in this series can be found on Nova's website under the E-book tab.

ENERGY SCIENCE, ENGINEERING AND TECHNOLOGY

SECURING GUAM'S ENERGY FUTURE

ASSESSMENTS, STRATEGIES AND ACTION PLANS

LUCIO N. BURROWS
EDITOR

nova publishers
New York

For permission to use material from this book please contact us:
Telephone 631-231-7269; Fax 631-231-8175
Web Site: http://www.novapublishers.com

Library of Congress Cataloging-in-Publication Data

ISBN: 978-1-62948-397-9

Published by Nova Science Publishers, Inc. † New York

CONTENTS

Preface **vii**

Chapter 1 Guam Initial Technical Assessment Report **1**
Ian Baring-Gould, Misty Conrad, Scott Haase, Eliza Hotchkiss and Peter McNutt

Chapter 2 Guam Strategic Energy Plan **105**
Misty Dawn Conrad

Chapter 3 Guam Energy Action Plan **135**
Misty Dawn Conrad and J. Erik Ness

Index **169**

PREFACE

This book provides an overview of various energy strategies that are available to Guam in the pursuit of diversifying fuel sources and reducing fossil energy consumption 20% by 2020 (20x20). The information in this book is based upon input from the Guam Energy Task Force (GETF), the National Renewable Energy Laboratory's (NREL's) Guam Initial Technical Assessment Report, and additional stakeholders. The information presented in this strategic energy plan is intended to be used by the GETF in developing an energy action plan. The next step in reaching the 2020 goal involves selecting the most appropriate strategies from the options outlined in this strategic plan and creating a roadmap for implementation in the Guam Energy Action Plan.

The strategies are categorized based on the time required to implement them. Near-term strategies can be accomplished in 1-2 years, mid-term strategies require 3-5 years, long-term strategies require 6-8 years, and some strategies require 8 or more years before they can be fully implemented. The prioritization of the strategies will help facilitate the development of an action plan and the selection of appropriate strategies for meeting Guam's 2020 goal.

Chapter 1 – On March 1, 2010, Assistant Secretary Tony Babauta from the Department of Interior's (DOI) Office of Insular Affairs (OIA), invited Governors Felix Camacho (Guam), Benigno Fitial (Commonwealth of the Northern Mariana Islands), and Togiola Tulafono (American Samoa) and their staff to meet with Senior Principals at the National Renewable Energy Laboratory (NREL) for discussions on ways to improve energy efficiency and increase the deployment of renewable energy technologies in the U.S. Pacific Territories. This meeting brought together major stakeholders to learn and understand the importance of developing a comprehensive strategic plan for implementing energy efficiency measures and renewable energy technologies.

For several generations, dependence on fossil fuels and the burden of oil prices have been a major concern. The urgency has never been more prevelant than today. With unstable oil prices, the volatility of supply and future military build-up plans on Guam, energy issues have become a high priority for the island. Succinctly, energy security is critical to Guam's future economic development and sustainability.

Under an interagency agreement, funded by OIA, NREL was tasked to deliver technical assistance to the island of Guam by conducting an island initial technical assessment that would lay out energy consumption and production data and establish a baseline. This assessment will be used to conduct future analysis and studies by NREL that will estimate energy efficiency and renewable energy potential for the island of Guam.

NREL provided an interdisciplinary team to cover each relevant technical area for energy assessments, system recovery, and upgrades. Experts in the following disciplines traveled to Guam for on-island site assessments October 11-15, 2010:

- Integrated Wind-Diesel Generation
- Transmission and Distribution
- Energy Efficiency and Building Technologies
- Solar Technologies
- Biomass and Waste-to-Energy
- Geothermal.

In addition to the above core disciplines, team capabilities also included expertise in program analysis, project financing, policy, and energy project, planning and Energy Committee facilitation.

The intent was to provide Guam with various scenarios and approaches to deploying cost effective energy efficiency and renewable energy technologies in a manner that will meet Guam's energy objectives. The information compiled in this energy assessment will be used as input in the development of a strategic plan.

Chapter 2 – This document presents various energy strategies that are available to Guam in the pursuit of diversifying fuel sources and reducing fossil energy consumption 20% by 2020 (20x20). The information in this document is based upon input from the Guam Energy Task Force (GETF), the National Renewable Energy Laboratory's (NREL's) *Guam Initial Technical Assessment Report*, and additional stakeholders. The information presented in this strategic energy plan is intended to be used by the GETF in developing an energy action plan. The next step in reaching the 20x20 goal involves selecting the most appropriate strategies from the options outlined in this strategic plan and creating a roadmap for implementation in the *Guam Energy Action Plan*.

The strategies are categorized based on the time required to implement them. Near-term strategies can be accomplished in 1-2 years, mid-term strategies require 3-5 years, long-term strategies require 6-8 years, and some strategies require 8 or more years before they can be fully implemented. The prioritization of the strategies will help facilitate the development of an action plan and the selection of appropriate strategies for meeting Guam's 20x20 goal. The strategies that will be used in reaching the 20x20 goal are broken down into five energy categories.

Chapter 3 – This document describes the four near-term strategies selected by the Guam Energy Task Force during action planning workshops conducted in March 2013, and outlines the steps being taken to implement those strategies.

Each strategy addresses one of the energy sectors identified in the Guam strategic energy plan as being an essential component of diversifying Guam's fuel sources and reducing fossil energy consumption 20% by 2020 (20x20).

The four energy strategies selected are:

1. Expanding public outreach on energy efficiency and conservation
2. Establishing a demand-side management revolving loan program
3. Exploring waste-to-energy options
4. Influencing the transportation sector: anti-idling legislation, vehicle registration fees, and electric vehicles

In: Securing Guam's Energy Future
Editor: Lucio N. Burrows
ISBN: 978-1-62948-397-9

Chapter 1

GUAM INITIAL TECHNICAL ASSESSMENT REPORT*

Ian Baring-Gould, Misty Conrad, Scott Haase, Eliza Hotchkiss and Peter McNutt

ABSTRACT

On March 1, 2010, Assistant Secretary Tony Babauta from the Department of Interior's (DOI) Office of Insular Affairs (OIA), invited Governors Felix Camacho (Guam), Benigno Fitial (Commonwealth of the Northern Mariana Islands), and Togiola Tulafono (American Samoa) and their staff to meet with Senior Principals at the National Renewable Energy Laboratory (NREL) for discussions on ways to improve energy efficiency and increase the deployment of renewable energy technologies in the U.S. Pacific Territories. This meeting brought together major stakeholders to learn and understand the importance of developing a comprehensive strategic plan for implementing energy efficiency measures and renewable energy technologies.

For several generations, dependence on fossil fuels and the burden of oil prices have been a major concern. The urgency has never been more prevelant than today. With unstable oil prices, the volatility of supply and future military build-up plans on Guam, energy issues have become a high

* This is an edited, reformatted and augmented version of National Renewable Energy Laboratory, Technical Report NREL/TP-7A40-50580, dated April 2011.

priority for the island. Succinctly, energy security is critical to Guam's future economic development and sustainability.

Under an interagency agreement, funded by OIA, NREL was tasked to deliver technical assistance to the island of Guam by conducting an island initial technical assessment that would lay out energy consumption and production data and establish a baseline. This assessment will be used to conduct future analysis and studies by NREL that will estimate energy efficiency and renewable energy potential for the island of Guam.

NREL provided an interdisciplinary team to cover each relevant technical area for energy assessments, system recovery, and upgrades. Experts in the following disciplines traveled to Guam for on-island site assessments October 11-15, 2010:

- Integrated Wind-Diesel Generation
- Transmission and Distribution
- Energy Efficiency and Building Technologies
- Solar Technologies
- Biomass and Waste-to-Energy
- Geothermal.

In addition to the above core disciplines, team capabilities also included expertise in program analysis, project financing, policy, and energy project, planning and Energy Committee facilitation.

The intent was to provide Guam with various scenarios and approaches to deploying cost effective energy efficiency and renewable energy technologies in a manner that will meet Guam's energy objectives. The information compiled in this energy assessment will be used as input in the development of a strategic plan.

Executive Summary

This document is an initial technical energy assessment for Guam, the first of many steps in developing a comprehensive energy strategy. The project plan for Guam includes three main objectives:

1. Assist in establishing an Energy Task Force
2. Publish an initial island assessment report to establish a baseline and identify energy efficiency and renewable energy opportunities
3. Support the Energy Task Force in developing a draft energy strategic plan that will provide Guam with various scenarios and approaches to deploying cost effective energy efficiency and renewable energy

technologies in a manner that will meet the island's objectives in reducing dependency on imported fossil fuels.

This report meets the second objective of establishing a baseline and identifying opportunities. A baseline gathers various data points to establish a reference against which progress can be measured. It allows comparison across sectors through trend identification and provides a tool for goal setting and measuring success of those goals. Before a baseline can be established, a methodology is developed.

The information compiled in this assessment will be used as input in the development of a strategic plan to meet the third project objective. This process sets the procedures and rationale that underlie the study.

This document summarizes data collected regarding energy production, consumption, and efficiencies, and discusses renewable energy and energy efficiency technology potential and current barriers and opportunities. The opportunities highlighted in this report can be used to formulate an energy plan. The following table highlights current energy efficiency and renewable energy opportunities designating them a low, medium, or high priority for impact.

Table 1. Energy Efficiency and Renewable Energy Opportunities and Potential Impacts

Opportunity Description	Impact Potential
Create a strategic plan to investigate and implement energy efficiency and renewable energy where technically appropriate and feasible	High
Increase energy efficiency standards in building codes	High
Further development of the Cotal 20 MW wind site with concurrent social acceptance outreach	High
Initiate a cool-roof program	Medium - High
Assess the potential for solar hot water heating in different sectors	Medium - High
Increase energy awareness through island campaigns	Medium - High
Continue to evaluate potential for sea water cooling project	Medium
Establish subcommittee to review options for possible modifications to PL 25- 175	Medium
DOD and GPA continue to work together to evaluate the geothermal potential	Medium
Set an energy efficiency standard for island appliances and air conditioning equipment	Medium
Outdoor lighting technology and control improvements	Low

Also within this report is a section that outlines barriers, possible next steps, and potential for technical assistance. Detailed technical information can be found in the appendices for further reference.

ACRONYMS AND ABBREVIATIONS

ASHRAE	American Society of Heating, Refrigerating and Air-Conditioning Engineers
bbl	billion barrels
Btu	British thermal unit
C&D	construction and demolition
CIA	Central Intelligence Agency
CT	Combustion Turbine
DEM	Digitial Elevation Model
DOD	Department of Defense
DOE	Department of Energy
DOI	Department of Interior
EIS	Environmental Impact Statement
EMD	Electro-Motive Diesel
EPA	Environmental Protection Agency
F	Farenheight
FY	Fiscal Year
GE	General Electric
GM	General Motors
GPA	Guam Power Authority
GRPP	Guam Resource Recovery Partners
GWA	Guam Waterworks Authority
GWh	gigawatt hour
HECO	Hawaiian Electric Company
HP	horsepower
IBM	Izu-Bonin-Mariana
IECC	International Energy Conservation Code
IFB	Invitation fore Multi-step Bid
IPP	Independent Power Producers
IRP	Integrated Resource Plan
km	kilometers
kV	kilovolt
kVA	kilovolt-ampere

kW	kilowatt
kWh	kilowatt hour
lb	pound
LED	Light-Emitting Diode
LEED	Leadership in Energy and Environmental Design
LFG	landfill gas
LIDAR	Light Detection and Ranging
m	meters
MECO	Maui Electric Company
MGD	millon gallons per day
mi	miles
m/s	meters per second
MMBtu	million British thermal units
MSW	municipal solid waste
MW	megawatt
MWh	megawatt hours
NAVFAC	Naval Facilities Engineering Command
NDWTP	Northern District Wasterwater Treatment Plant
NOx	nitric oxide
NRCS	National Resources Conservation Service
NREL	National Renewable Energy Laboratory
OIA	Office of Insular Affairs
PL	Public Law
PMC	Performance Management Contract
PUC	Public Utilities Commission
RFO	residual fuel oil
RFP	Request for Proposal
RPM	revolutions per minute
RPS	Renewable Portfolio Standard
SARs	Severe Acute Respiratory Syndrome
Sf	square foot
SWMD	solid waste management district
USDA	U.S. Department of Agriculture
WTE	waste-to-energy

1. METHODOLOGY

To clarify the National Renewable Energy Laboratory's (NREL) team's understanding of the project, the approach and background information used for the assessment have been outlined in Sections 1.1 and 1.2.

1.1. Approach

Establishing the parameters of a study provides focus for the assessment team. Certain topic areas, such as transportation, are considered important to overall energy strategies, however were not considered within the Guam assessment due to budget constraints and the priority of targeting energy efficiency and renewable energy technologies.

Each assessment begins with a background study to understand the site demographics and geography. Human and physical geography impact energy consumption and the success of implementation of renewable energy projects, therefore this information is essential to the types of recommendations made.

Data collection and researching previous studies allows the assessment team to understand what the current baseline consumption is and what recommendations have been made in the past. The data collection process coincides with a site visit to interview stakeholders and survey the options to be included within the assessment. After the site visit, information is analyzed and opportunities are identified.

The graphic in Figure 1 illustrates the major components of NREL's assessment methodology.

It is essential to look at the energy picture of Guam holistically. The Department of Defense (DOD) has considerable representation on-island, accounting for approximately 21% to 22% of Guam's energy consumption. With an impending military build-up, concerns surrounding energy use are receiving increased attention. An extensive understanding of the relationship between Guam's military installations and the civilian side of the island is necessary to establish energy efficiency and renewable energy baselines from a whole island approach. NREL's assessment team concentrated on baseline factors that accounted for DOD presence.

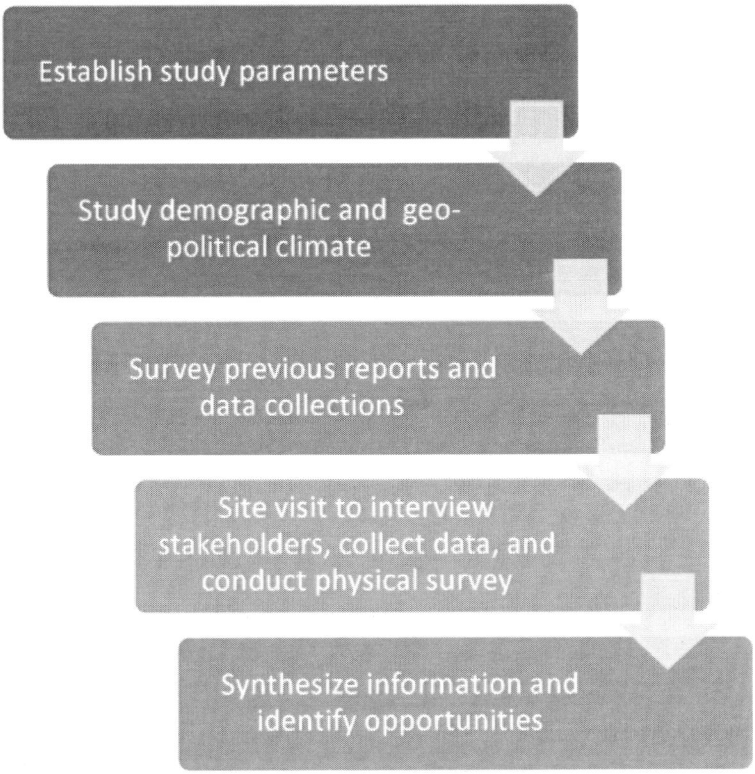

Figure 1. Assessment methodology.

This report focuses on island-wide utility generation and consumption patterns, not specific DOD energy efficiency and renewable energy activities. DOD has its own internal goals for renewable energy and energy efficiency. As DOD seeks to meet these goals, there may be impacts on the operations and economics of Guam Power Authority (GPA)—it is important for DOD and GPA to continue to work together to understand these impacts and seek to develop win-win solutions.

There are many energy efficiency and renewable energy technologies on the market today. Included within this study's parameters are mature technologies that are considered prime time and bankable. There are other technologies that Guam may want to investigate for future study, but for the basis of this study, technologies included had the following criteria:

- Commercial availability
  - Tested and demonstrated
  - Carry warranty
  - Service and parts available
- Ready for immediate deployment
- Sound investment
- Financing available from private sector organizations.

This assessment focuses on clean energy technologies. There is varied opinion on what falls under this purview. For the purposes of this study, commercial clean energy technologies were identified under the current NREL portfolio, including wind, solar, biomass, energy efficiency, and geothermal. NREL did not assess opportunities associated with ocean thermal energy conversion, off-shore wind, marine hydrokinetic, nuclear, coal, or liquefied natural gas.

This document summarizes data collected regarding energy production, consumption, and efficiencies, and discusses renewable energy and energy efficiency technology potential and current policy barriers and opportunities.

1.2. Background

Guam is an island in the North Pacific Ocean (13° north and 144° east). Guam is an unincorporated territory of the United States under the jurisdiction of the Office of Insular Affairs (OIA). Guam has a total land area of 544 km^2 (210 mi^2) with 125 km (130 mi) of coastline, making it the largest and southernmost island in the Mariana Archipelago. Figure 2 provides a basic map of Guam[1].

The island is volcanic, has coral reefs, a coralline limestone plateau, which is the source of most of the island's fresh water; and cliffs along the coast and mountains in the south of the island (the highest point is Mount Lamlam at 406 m). Permanent crops cover 18% of the island. Natural hazards include squalls during the rainy season, typhoons (between June-December), and invasive predators like the brown tree snake that decimated the local bird population and the Rhinoceros beetle, which is destroying the coconut palm.

In 2000, the island's population was 154,805[2] and the 2010 estimate is 180,865[3] , spread throughout about 20 villages. In 2000, the five largest villages were Dededo (42,980), Yigo (19,474), Tamuning (18,012), Mangilao (13,313), and Barrigada (8,652). According to the OIA website, the population in 2010 was 178,000, which is slightly lower than the Central Intelligence Agency estimates of 180,865. The value of 180,865 is assumed in this report when making any per capita estimates.

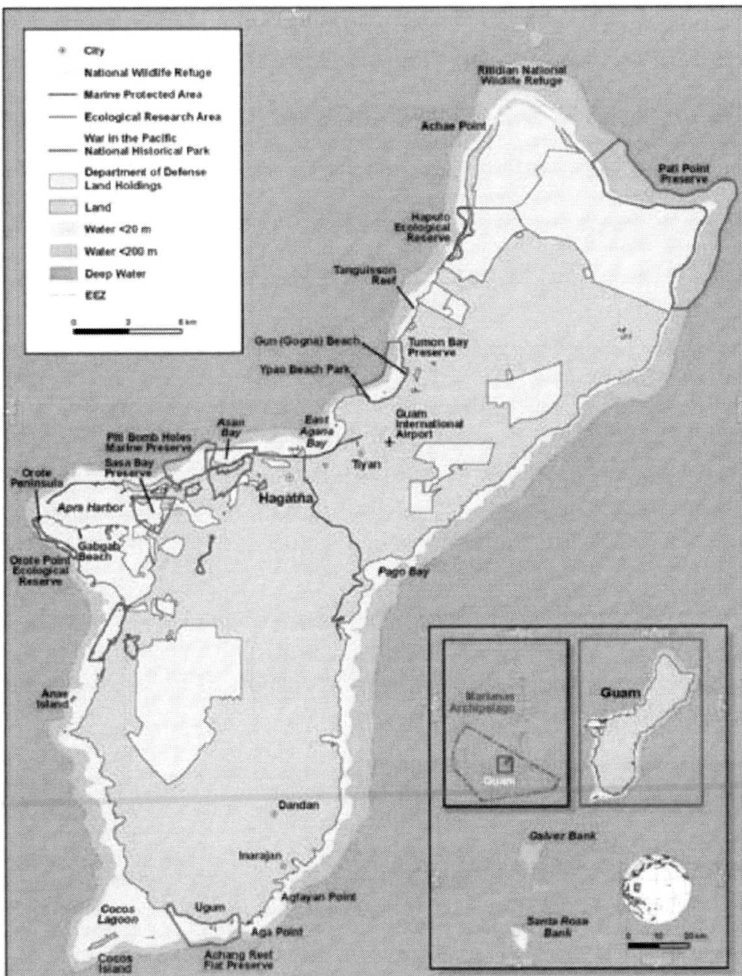

Source: NOAA.

Figure 2. Map of Guam.

With the planned military buildup (relocating troops from Okinawa to Guam) these figures are expected to grow fairly significantly, but are always changing. There are many factors to consider; funding availability, infrastructure and construction readiness, 2011 earthquake and tsunami in Japan, etc. Table 148 in Appendix C provides additional information provided from the Environmental Impact Statement (EIS).

The military build-up will impact the electricity consumption and generation needs on Guam. As the surge of military activity increases, economical impacts are expected. The estimated expected population is summarized in Table 2.[4]

Table 2. Estimated Additional Military and Associated Population in 2017

Marines moving from Okinawa	Dependents	Civilian Employees	Temporary Workers	Estimated Cost
8,600	~9,000	6,000	~20,000	$10.27 billion

Cultural dynamics often impact energy consumption as well. Guam is rich with diversity; ethnic groups include native Chamorro (37.1%), Filipino (26.3%), Pacific Islanders (11.3%), white (6.9%), other Asian (6.3%), other ethnic origins (2.3%), and mixed ethnicities (9.8%).[5] The predominant religion is Roman Catholic. Languages include English (38%), Chamorro (22%), Filipino (22%), Pacific island languages (6%), and Asian languages (7%). Guam has a literacy rate of 99%.

The economy is driven mostly by military and tourism; the tourism industry has grown since the 1980s due to a push to attract Japanese tourists. With a total of 1-1.2 million tourists a year, Japanese tourists made up 76% of tourists in 2007 with direct tourist spending being $1.2 billion. The basic breakdown of total payroll employment and percentage of total is summarized in Table 3.

Table 3. Employment Statistics[6]

Total payroll employment (2007)	60,350
Private sector employment	45,320
Government employment	15,030
Private sector employment as a % of total	75.1%
Government employment as a % of total	24.9%
Unemployment rate	11.4%
Private sector average weekly earning	$411.55
Government of Guam average weekly earnings	$737.48

Guam has a tropical marine climate that is warm and humid with little seasonal temperature variation, illustrated in Figure 3. Climate impacts the energy consumption in a specific geographic region.

Source: NREL/PIX 18625.

Figure 3. War in the Pacific Park.

Table 4. Guam Historic Weather Summary[7]

Agana, Guam

Elevation: 246 feet Latitude: 13 29N Longitude: 144 48E

Average Temperature Years on Record: 47

	YEAR	Jan.	Feb.	Mar.	Apr.	May	Jun.	Jul.	Aug.	Sep.	Oct.	Nov.	Dec.
°F	81	80	80	81	82	82	83	82	82	82	82	82	81

Average High Temperature Years on Record: 47

	YEAR	Jan.	Feb.	Mar.	Apr.	May	Jun.	Jul.	Aug.	Sep.	Oct.	Nov.	Dec.
°F	86	85	85	86	87	87	88	87	87	87	87	87	86

Average Low Temperature Years on Record: 47

	YEAR	Jan.	Feb.	Mar.	Apr.	May	Jun.	Jul.	Aug.	Sep.	Oct.	Nov.	Dec.
°F	76	75	74	75	76	77	77	76	76	76	76	77	76

Highest Recorded Temperature Years on Record: 47

	YEAR	Jan.	Feb.	Mar.	Apr.	May	Jun.	Jul.	Aug.	Sep.	Oct.	Nov.	Dec.
°F	96	94	93	93	96	94	95	95	93	93	93	92	91

Lowest Recorded Temperature Years on Record: 47

	YEAR	Jan.	Feb.	Mar.	Apr.	May	Jun.	Jul.	Aug.	Sep.	Oct.	Nov.	Dec.
°F	65	67	65	66	68	70	70	70	70	70	67	68	68

Average Precipitation Years on Record: 47

	YEAR	Jan.	Feb.	Mar.	Apr.	May	Jun.	Jul.	Aug.	Sep.	Oct.	Nov.	Dec.
in.	86.8	4.4	3.2	2.5	3.4	5.3	5.6	9.6	13.4	13.1	12.5	8.5	5.1

The U.S. Department of Energy (DOE) has issued maps of various climate zones based on temperature and humidity. Guam's climate is tropical marine with the dry season between January-June and the rainy season between July-December. Yearly average temperatures range between 76°F and 86°F and average annual precipitation is 86 inches. Table 4 summarizes the historic weather on the island over the past 47 years.

2. Establishing a Baseline

Following a review of the natural and cultural influences, site assessments incorporate historical energy consumption data to establish the baseline. This section summarizes observations from electrical generation, end-use energy consumption and costs, and solid waste management. Additional details are provided in the appendices.

2.1. Overall Fossil Fuel Energy Use

With the exception of a few small renewable energy projects, Guam is 100% dependent on fossil fuels for meeting its energy supply needs. The primary sectors that consume fossil fuels in Guam are:

- Production of electricity
- Transportation
- Water heating and self-generation by the commercial sector (primarily hotels)
- Propane for home cooking and small commercial operations.

NREL obtained information on fossil fuel consumption for electricity production, transportation, and propane use by consumers. We did not attempt to locate data on consumption by hotels or small commercial sectors. In 2010, GPA consumed nearly 2.9 million barrels of fossil fuel (120,519,396 gallons) for power generation.[8] The Guam Energy Office collects data on fuel sales, including jet fuel and consumer propane.[9]

In 2010, 109 million gallons of motor fuel were sold, including gasoline, diesel, and jet fuel. Note the motor fuel data compiled by the Guam Energy Office does not include shipping to Guam, air transport and freight and DOD.

In 2010, Guam used 230 million gallons of fossil fuel, or over 1,270 gallons per year for every man, woman, and child living on the island. This value can serve as the primary baseline for setting fossil fuel reduction targets and against which future program impacts can be measured. Table 5 shows the breakdown by end use, and Figure 4 shows these values graphically.

Table 5. 2010 Guam Fossil Fuel Usage

End Use	Gallons/year
Gasoline	46,262,100
No 2 Diesel	14,105,600
Jet	44,812,200
Propane - Consumer	4,305,500
GPA	120,519,396
Total Fuel Used	230,004,796
Population	180,865
Fuel use per capita	1,272

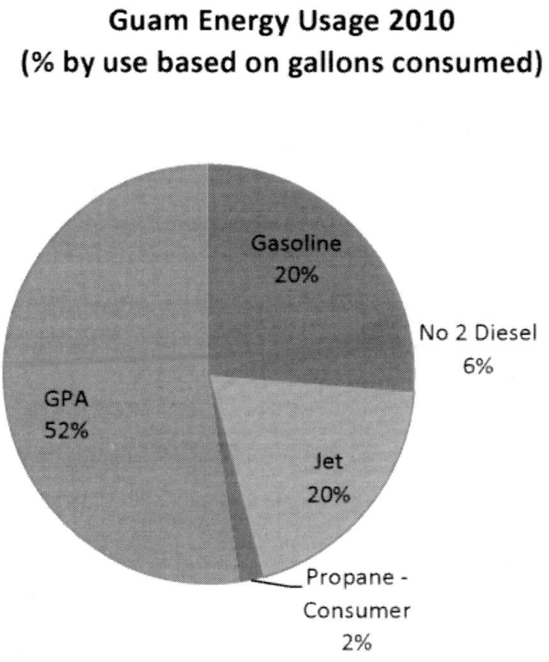

Source: GPA.

Figure 4. Total fossil fuel energy use, 2010.

2.2. Electrical Energy

2.2.1. Electrical Generation

GPA is a full service electric utility serving about 46,000 customers. GPA owns and manages 663 miles of transmission and distribution lines and 29 substations throughout the island. Electricity is delivered through a transmission and distribution grid consisting of a combination of 115 kV, 34.5 kV, and 13.8 kV lines. Its generation capacity is 552 MW with total energy production of 1,854 GWh (FY 2009). The highest peak system demand was 281.5 MW in June 2001.

GPA's generation includes 181 MW from Independent Power Producers. The generation mix is 33% steam turbine, 30% slow-speed diesel, 26% combustion turbine, and 10% fast-track diesels. GPA has organized 210 MW of its base load capacity under two performance management contracts (PMCs). These PMCs provide private management using public employees to operate and maintain the plants. The contracts contain performance-based incentives for reducing plant operating costs. The major fuel types include residual fuel oil (RFO #6) and diesel #2 distillate. Figure 5 shows GPA's generation by fuel type. Note that in FY 2010 99% of GPA's electricity generation came from RFO. RFO includes both high and low sulfur sources.

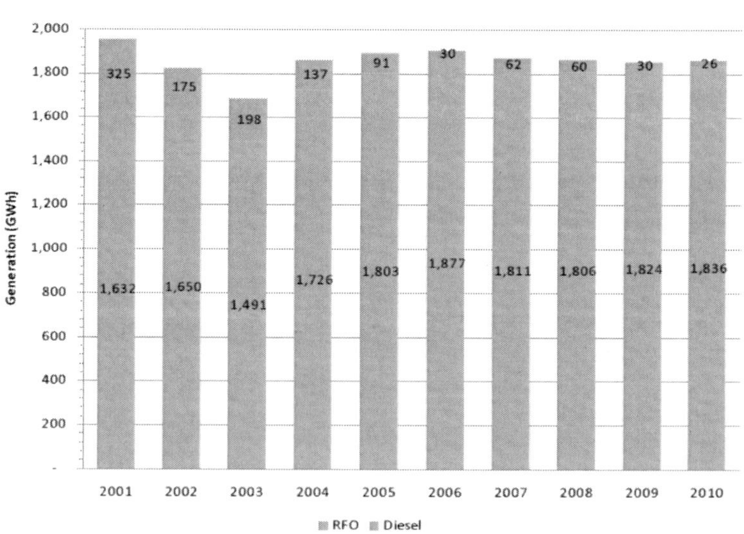

Source: GPA.

Figure 5. CPA gross generation by fuel type.[10]

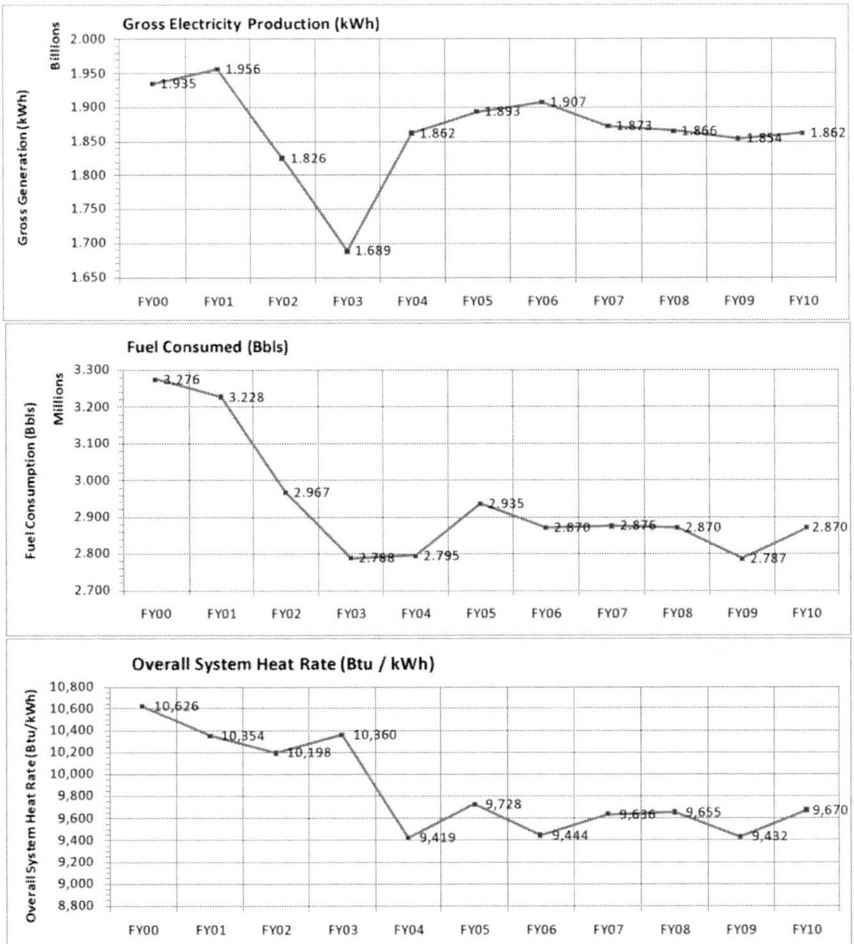

Source: GPA.

Figure 6. CPA gross electricity production, fuel consumption and overall system heat rate.[15]

According to GPA, "several systemic shocks, such as the Gulf war, SARs [Severe Acute Respiratory Sydrome], major typhoons, and rising fuel prices occurred during this period, adversely affecting the Guam economy. As a result, the demand for electric power contracted or remained flat relative to FY 1998," (Guam Power Authority, 2008).

The top plot in Figure 6 shows the gradual decline of generation, from a high of 1.956 billion kWh in 2001 to 1.862 billion kWh in 2010, a 5% decline in nine years[11] . Electricity generation is measured at the generators in the power plants. The fuel use also declined during this same period, while the fuel efficiency improved between 2003 and 2004 and remained relatively flat during the last seven years.

In 2010 the overall system heat rate was 9,670 Btu/kWh. GPA provided historical fuel heating values from which the average value of 6.275 MMBtu/bbl for high sulfur fuel oil and low sulfur fuel oil 12. Although the fuel heating value of diesel is value is 5.760 MMBtu/bbl, because diesel is only a small proportion of the fuel mixture, it was not broken out separately. As a comparison to other island power systems, Hawaii's overall system heat rate was about 10,500 Btu/kWh[13] and the U.S. Virgin Island's system-wide average was about 15,200 Btu/kWh[14].

2.2.2. Consumption and Costs

Figure 7 shows the overall annual energy consumption on Guam. The shape of the curve corresponds with the gross generation curve (top plot of Figure 6), showing a 6% decline in consumption between 2001 and 2010. The difference between generation and consumption is attributed to power-plant auxiliary losses, line losses, and losses from technical (billing errors, broken meters) and non-technical (theft) sources. Total losses in 2010 were about 12%. There is a drop in 2003 due to several tragic events; Typhoons Chata'an, Pongsonga, and Halong all impacted Guam in 2002 causing extensive damage to infrastructure. Additioanlly, SARS and the attacks of September 11, 2001, led to reduction in tourism.

Figure 8 shows three daily load profiles for the overall GPA system. The hourly data are taken over the one-year time period between March 1, 2009 and February 28, 2010. The green plot is the hourly average over the entire year. The red plot shows hourly data for the day of the year in which the hourly maximum load occurs (October 13, 2010), and the blue plot shows hourly data for the day of the year where the hourly minimum load occurs (May 7, 2009). On an annual average basis, the minimum load (approximately 180 MW) occurs between 04:00 and 05:00, and the peak load (around 240 MW) occurs between 19:00 and 20:00, representing a 25% change in load. On the day with the hourly maximum load, the minimum of 189 MW occurs at 04:00, and the maximum of 272 MW occurs at 19:00, representing a change in load of 34%. On the day with the hourly minimum load, the minimum of 144 MW occurs at 01:00, and the maximum of 247 MW occurs at 20:00, representing a change in load of 42%.

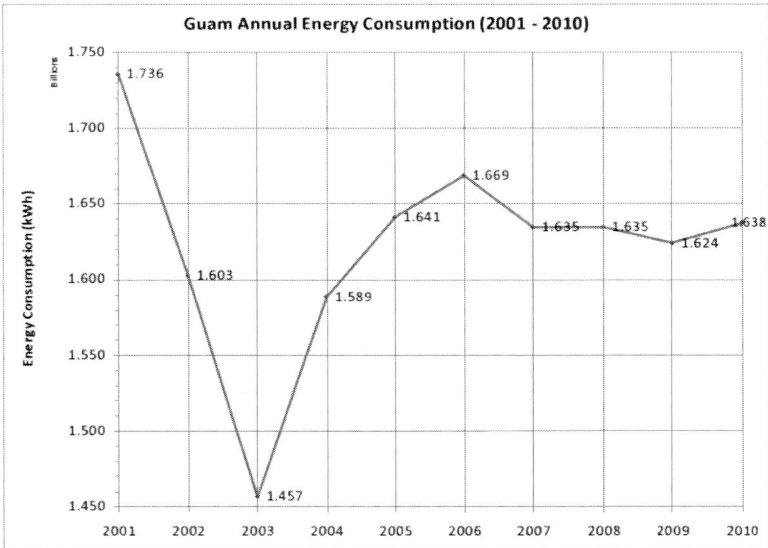

Source: GPA.

Figure 7. Historical annual island energy consumption (2001 – 2010).[16]

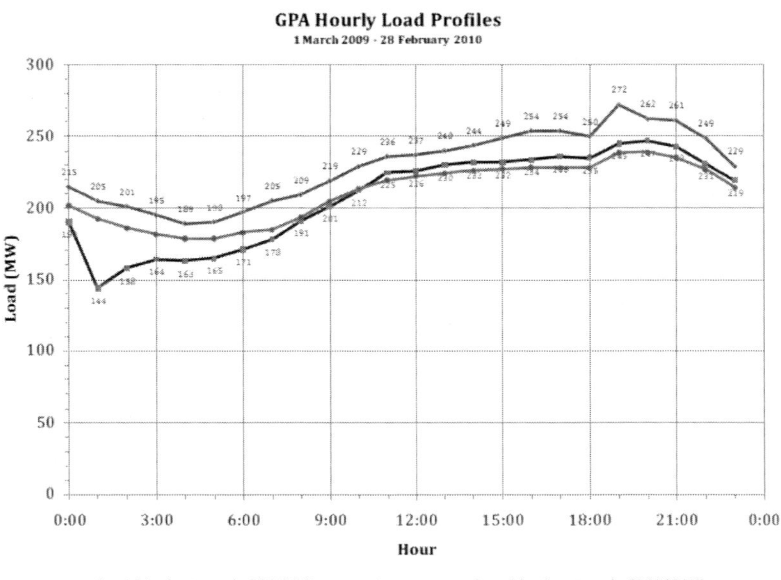

Source: GPA.

Figure 8. Overall CPA system hourly load profiles (March 1, 2009 – February 28, 2010).[17]

Electricity is produced by imported oil (RFO #6) and diesel (#2 distillate). The total consumption of oil in 2007 was 14,230 bbl/day[18]. In 2009, the average cost of electricity was $0.23 per kWh and the GPA's 2010 Levelized Energy Adjustment Clause factor was $0.12465/kWh (Effective August 1, 2010)[19]

In order to understand where energy efficiency measures should be focused, the electricity consumption has been broken down by end-consumer (see Figure 9). The energy statistics available for the island present energy consumption by commercial, residential, U.S. Navy, government, and street lighting sectors. Historical data is available for the period between 2001 and 2010, so a historical average was used to compare these sectors over the 10-year timeframe[20]. The largest consumers of electricity on Guam are the commercial and residential sectors, followed by the U.S. Navy and government. As a group, private hotels dominate GPA's commercial load, and the airport is the largest civilian critical load. Figure 10 shows the Guam energy consumption by sector between 2001 and 2010.

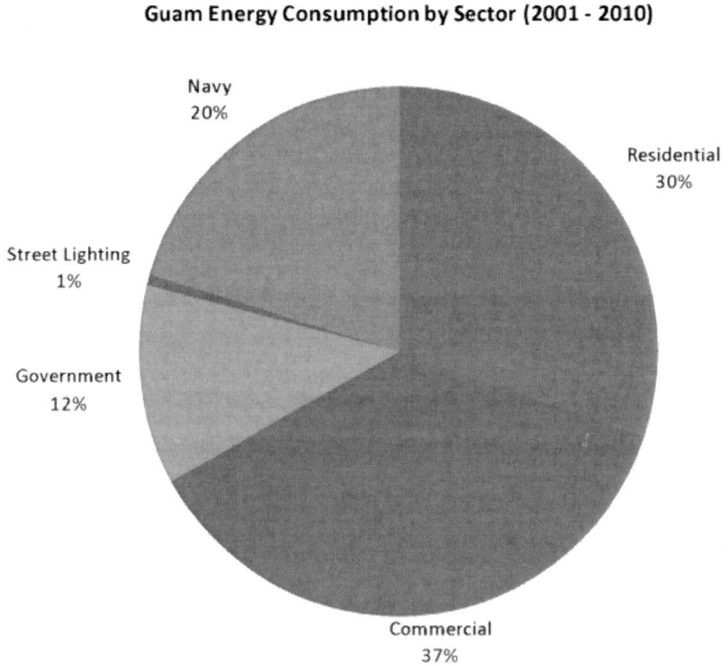

Source: GPA.

Figure 9. Guam energy consumption by sector (2001 – 2010 average).[21]

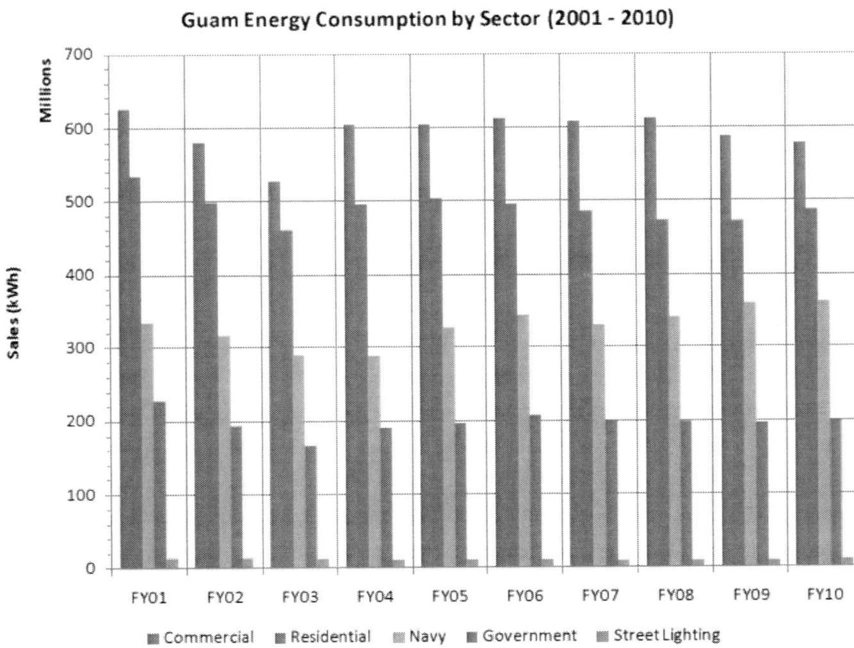

Source: GPA.

Figure 10. Guam energy consumption by sector (2001 – 2010).[22]

GPA is working with DOD during planning to ensure there is adequate capacity to serve the load of military projects targeted for the island. The current best estimate of the additional base load from the buildup is 30 MW with an additional 25 MW of transient/peaking load. GPA believes there is sufficient generation capacity currently in the system to meet the planned load requirements. DOD has also expressed that a preferred solution to complement GPA's existing generation system is to recondition up to five peaking/emergency combustion turbine units. GPA is working with DOD to identify transmission and distribution projects that would be required to meet the military service requirements. DOD has been committed to the principle of ensuring the civilian community is not negatively impacted by the cost of the buildup.

2.2.3. End Use

Understanding the building stock is important to know how energy is being consumed within a building, as energy use will vary by construction type/materials and the age of the building. Other parameters that impact energy use are climate and culture. The goal in reducing energy consumption is to understand how it is being utilized and to know where conservation and efficiency measures can be applied. By identifying the end-uses of energy consumption, analysis can be undertaken to identify areas where improvements can be made.

NREL's site visit demonstrated that in typhoon prone areas, such as Guam, buildings are typically constructed of concrete to withstand the wind and water. Buildings that are lightweight and constructed of wood frames or are older and have deteriorating structures are vulnerable to typhoon destruction.

Reinforced concrete is often used as the building material in typhoon 'alleys'; this type of construction was observed in Guam. The building portfolio is comprised of hotels, the airport, government and military buildings, schools, hospitals, medical clinics, commercial property, and residential buildings, such as apartments and single family dwellings. The number of buildings classified within each category is unknown; however opportunities were identified within the energy end-use sectors as provided by the Government of Guam and GPA. The direction the Government is trying to head is exemplified by the Leadership in Energy and Envionremental Design (LEED)-registered buildings under construction on the Guam Community College Campus, shown in Figure 11.

Source: NREL/PIX 18588.

Figure 11. Guam Community College's LEED-registered building under construction.

Data collected from the site visit have indicated that the highest end users of energy on Guam are commercial, residential, government, Navy, and then street lighting, as shown in the pie chart above in Figure 9. End use within the commercial, residential, and government sectors are discussed in more detail in the following sections. Delving into the end use in more detail allows for solutions to be identified either in terms of energy conservation or energy efficiency.

Energy conservation and energy efficiency can be combined to create an effective strategy to achieving overall island energy reduction. **Energy conservation** is any behavior that results in the use of less energy. Turning the lights off when you leave the room is one example of conserving energy. **Energy efficiency**, on the other hand, is the use of technology that requires less energy to perform the same function. A compact fluorescent light bulb that uses less energy than an incandescent bulb to produce the same amount of light is an example of energy efficiency[23].

2.2.3.1. End Use for Commercial Sector

Commercial building consumption is led by offices, retail, luxury hotels, and the education sector. End users within commercial buildings are reported by GPA as being interior lighting, cooling, miscellaneous equipment, refrigeration, ventilation, exterior lighting, water heating, and cooking. The end use is broken out for commercial buildings in Figure 12[24]. 1993 is the most current end use data available.

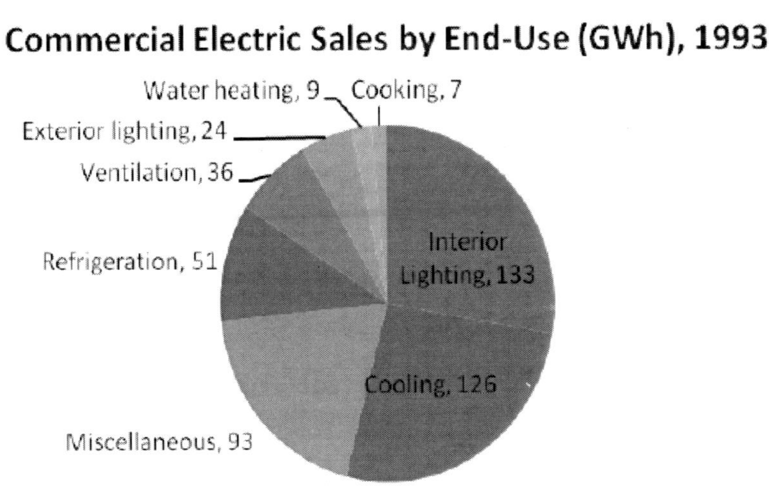

Source: GPA.

Figure 12. Commercial electric sales by end-use, 1993.

Many assumptions were made by the authors commissioned to calaculate the sector break-down for GPA; therefore some scrutiny may be necessary when analyzing these results. A more detailed study may be required in commercial buildings to determine whether interior lighting is truly higher than cooling.

Additionally, more information on the end-users within the 'miscellaneous' category would be helpful. By understanding more about the equipment and plug loads that are consuming energy, a commercial organization will have better control over managing and reducing their energy consumption. An example is a company that leaves computers, printers, and fax machines on or on standby overnight. Without knowing what the components of the 'miscellaneous' category are it is difficult for the organization to know how to make reductions.

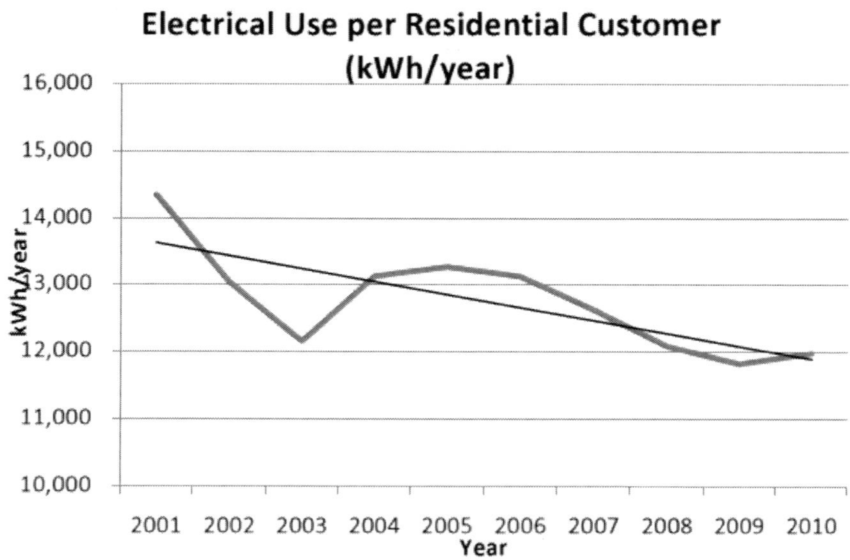

Figure 13. Annual energy sales (residential sector).

2.2.3.2. End Use for Residential Sector

GPA reports that consumption by the residential sector is impacted by the rate of electricity, which fluctuates primarily based on the cost of fuel. Despite a growth in population and in electricity customers, Figure 13 shows that the average kWh sales per residential customer declined by 16% between 2001 and 2010[25]. This decrease is thought primarily to reflect reduced consumption due to higher rates (shown in Figure 14). From 2001 through 2010, the average annual fuel bill for GPA's residential customers has increased by 44%. The graphs show the actual activity in the thicker, blue line; the average trend lines are shown in the thin, black line.

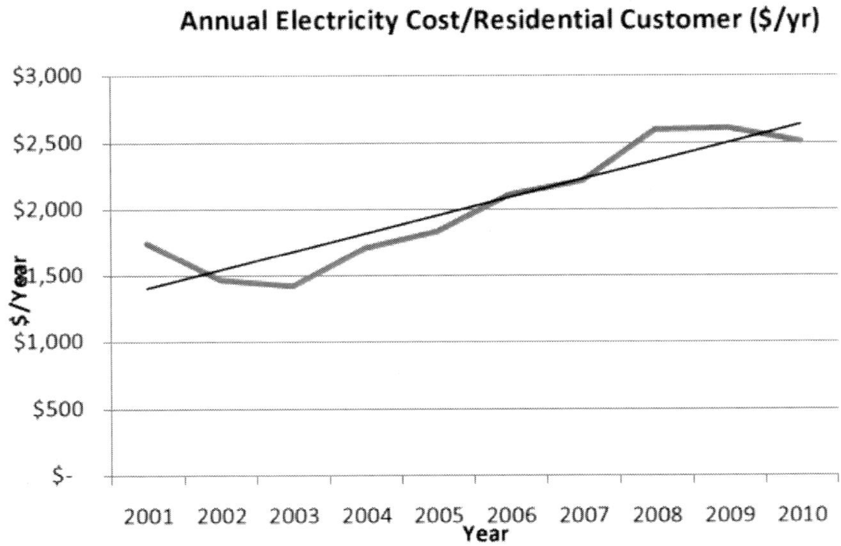

Source: GPA.

Figure 14. Average annual costs per residential customer.

In 1993, the highest energy consumption in Guam within the residential sector was water heating, followed by air conditioning and household appliances, as shown by the end use of energy in residential buildings shown in Figure 1526. This tells us the best opportunities for the residential sector include reducing electrical use for water heating, improving air conditioning, improving appliances, and changing to more efficient lighting.

2.2.3.3. End Use for Government Sector

A snapshot glimpse for the Guam Government sector indicates that the largest consumption of fuel for November 2010 was related to a few specific agencies. The Guam Waterworks Authority (GWA) had the largest energy consumption, followed by the Department of Education, the University of Guam, and the hospital. This breakdown is shown in Figure 16.

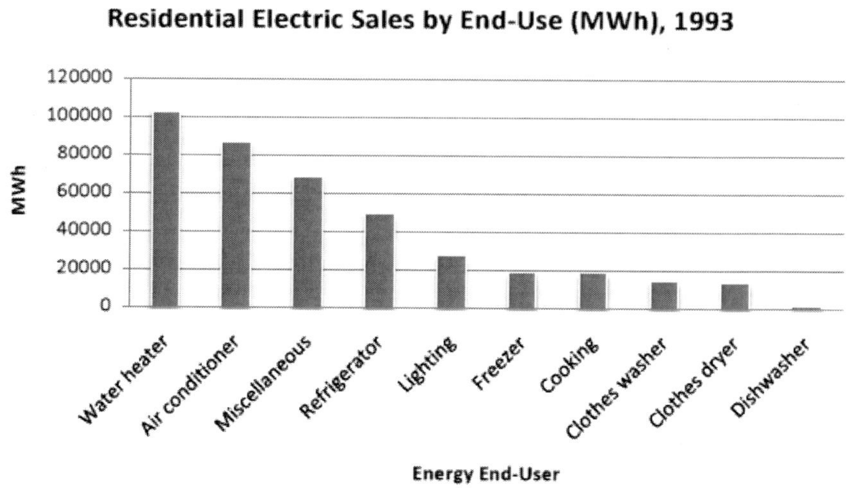

Soure: NREL.

Figure 15. Residential electric sales by end user.

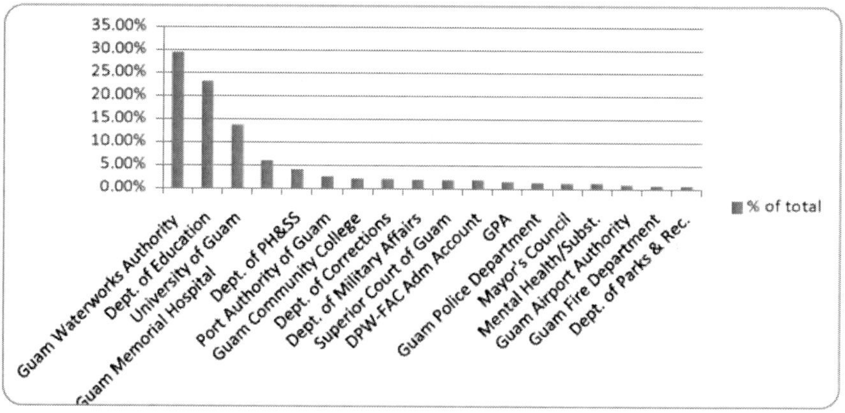

Source: NREL.

Figure 16. Guam government sector energy consumption by percentage of total energy consumption.[27]

While this type of assessment is not comprehensive in its review of energy consumption, it provides a glimpse of where energy is being consumed within the sector. Ideally, an assessment would cover multiple years worth of data and be compared to the total floor area of each agency. The energy consumption per floor area provides a more comprehensive form of benchmarking energy consumption activities within buildings and highlights where energy is being consumed more thoroughly.

Without further investigation it is difficult to confirm assumptions; however assumptions can be made about the end use within each agency. For example, the energy consumption at GWA may be a result of typical energy end use, such as lighting, air conditioning, and equipment; however motors, drives, and pumps may contribute significantly to the consumption. GWA is responsible for treating and pumping water, therefore it can be assumed that the primary energy consumers are due to these main activities. The Department of Education, the University of Guam, and Guam Community College are most likely responsible for a large building portfolio with high cooling, lighting, and plug loads, whereas the hospital presumably has the same loads, but on a 24-hour demand scale.

2.2.4. Current Status of Energy Policy

Across the global landscape, many communities are turning to policy as a way to institute clean energy technologies. Many states, territories, local governments, and communities are interested in adopting strategies that promote energy efficiency and renewable energy. Policy is a mechanism many governments use to advance and support energy efficiency and renewable energy objectives. Policy implementation can aid in market adoption and assist in laying a foundation for a green economy while providing investors with reassurance within market transformation. Policy can be an effective tool Guam can use to implement their energy efficiency and renewable energy goals, increase energy diversity, and decrease dependence on imported fossil fuels. Additionally, there are environmental benefits such as emissions reduction, reduction of water consumption, thermal pollution and adverse land-use impacts. Oftentimes, these environmental effects can be beneficial in a cost-benefit analysis when factoring in future fossil fuel generation.

The following subsections discuss energy policies currently in place in Guam.

2.2.4.1. Net Metering

Net metering for distributed generation was enacted in December 2004 under Guam Public Law (PL) 27-132.[28] The net metering policy is an agreement between GPA and its customers for a variety of renewable technologies; fuel cells, microturbines, wind energy, biomass, hydroelectric, solar, or hybrid solar systems. In order to participate in this program the system cannot exceed the eligible size of 25 kW. GPA will pay for any additional meters needed as well provide customer-generators with a bi-directional energy meter. Customers must meet all prescribed safety guidelines and electric codes. GPA does not require additional measures, controls, or supplemental insurance once these requirements have been met. Customers who take advantage of net metering can request billing on either a monthly, semiannual, or annual basis. Systems that produce excess generation will be compensated; however, the rules for this policy have yet to be set by the Guam Public Utilities Commission (PUC), which was required within 120 days of the authorizing legislation.

Through interviews it has been learned that the PUC has currently enacted these rules. If the energy the customer produces is zero or negative during a particular billing cycle, the customer shall pay only the non-energy charge portion; if the customer's net energy is negative they will be credited the kilowatt hour different in the next billing cycle.[29] Table 6 lists the current net metering customers.

Table 6. Current GPA Net Metering Customers

Location	Technology	Size (KW)	Cust Type	Activation Date
Golondrina Avenue, Barrigada Heights	PV	8.4 KW	Residential	2/17/2009
Flores Rosa Loop,Barrigada Heights	PV	8.4 KW	Residential	5/19/2009
Daisy Lane,Latte Heights	PV	7.7 KW	Residential	4/17/2009
Tun Emeterian Camacho St.,Oka,Tam.	PV	8.4 KW	Residential	9/25/2009
Acho Latte St.,Macheche,Dededo(2 ea.)	PV	7.7 KW (2 ea.)	Commercial	1/22/2010
Toto Canada Rd.,Barrigada	PV	2.96 KW	Residential	12/1/2009
Toto Canada Rd.,Barrigada	Wind	1.8KW	Residential	6/22/2009
Chalan Villagomez,Latte Heights	Wind	1.8KW	Commercial	6/1/2009
Mangilao , GCC Compound, New Learning Center	PV	24.8KW	Commercial	12/10/2010
Mangilao , GCC Compound, Allied Health Center	PV	29.4KW	Commercial	Under Construction

2.2.4.2. Energy-Efficient Appliance Rebate Program

The "State Rebate Program" sponsored by DOE through the American Recovery and Reinvestment Act of 2009 rolled out in Guam 2010[30] . Unfortunately, funds are limited and this program has not been locally invested and therefore will have a short lifespan. However, incentives to replace inefficient equipment with new ENERGY STAR certified products, including washing machines, refrigerators, and air conditioners (both room and split type models), will no doubt show measureable savings. Approximately $160,000 was set aside for this program.

2.2.4.3. Solar-Ready Residential Building Requirement

Effective in October 2000, the Guam Energy Code adopted installation requirements for new residential low-rise buildings. The building code now requires that all new residential construction must include the appropriate solar thermal equipment that would allow future solar thermal unit installation[31] .

2.2.4.4. Building Codes

The current building codes in Guam are outdated. The Guam Energy Code uses ASHRAE 90.1-1989 for commercial buildings and follows the 1992 Model Energy Code for residential buildings.[32] These codes were adopted more than 12 years ago with provisions to allow the Department of Public Works to administer the latest edition of the code for design and construction.

The most current green building codes to be recommended are:

- **Commercial:** ASHRAE 90.1 2010, Energy Standard for Buildings Except Low-Rise Residential Buildings/International Energy Conservation Code (IECC), 2009
- **Residential:** International Energy Conservation Code, 2009.

These codes are set up to address heat loads, but tropical areas can choose what is applicable as the IECC tends to group energy requirements and discount tropical climate impacts on buildings.

As of October 28, 2010, Guam adopted the 2009 IECC. Part of the requirements of that adoption are to establish a Guam Building Code Council whose responsibilities will include providing "recommendations for further amending or adding to the codes adopted with this legislation, including the possible amending or adoption of Appendices A, B, and D of the International Building Code, 2009 edition, the International Property Maintenance Code, 2009 edition, the International Energy Conservation Code, 2009 edition, the Model Tropical Energy Code, and any other construction codes as the Council may deem appropriate,"[33] . Adoption of the Tropical Building Code is only the first step to addressing energy use in new construction and major renovations; but current building stock is a concern. Compliance and enforcement are other issues that must be addressed.

2.2.4.5. Renewable Portfolio Standard

Guam has not adopted the traditional Renewable Portfolio Standard (RPS) but has an RPS goal that was introduced as Guam Bill 166 and enacted in March 2008 through the Guam legislature—Guam Public Law 29-62[34] . This law was an attempt to address three important initiatives: (1) promote the development of renewable energy, (2) establish an RPS, and (3) study the feasibility of a rate structure through the PUC that would support renewable energy implementation. It established a renewable energy goal of 25% renewable energy by 2035. This goal is not compulsory like a traditional RPS, which is legally binding.

If Guam is to move toward decreased fossil fuel use and increased energy diversity, an RPS can be a very effective tool. The following is Guam's timeline for renewable energy integration as spelled out in their 2008 RPS goal.

- 5% of net electricity sales by December 31, 2015
- 8% of net electricity sales by December 31, 2020
- 10% of net electricity sales by December 31, 2025
- 15% of net electricity sales by December 31, 2030
- 25% of net electricity sales by December 31, 2035.

Here are a couple island examples:

U.S. Virgin Islands has passed legislation and established that "peak demanded generating capacity" come from a renewable source of 30% by 2025.[35]

Hawaii has established a 70% fossil fuel reduction target by the year 2030, 40% of its net electricity sales.

These goals are all inclusive of energy usage as a whole and include the transportation sector.[36]

2.2.5. Solid Waste

Solid waste generation and disposal represent significant costs and challenges for island communities like Guam. Solid waste represents a potential source of material that can be converted to energy. The first step in determining the potential is to understand present waste management practices, waste generation quantities, waste composition, disposal costs, and regulatory environment.

Solid waste management issues in Guam are complicated. Under a Consent Decree issued in 2004, the United States District Court ordered the Government of Guam to plan for the closure of Ordot Dump and to construct a new sanitary landfill that would be fully compliant with subtitle D of the Resource Conservation and Recovery Act[37].

Over the next four years, the Government of Guam made little progress in implementing the Consent Decree. In March 2008, the District Court of Guam issued a Court Order[38] to appoint consulting firm Gershman, Brickner and Brayton (GBB) as the Federal Receiver to enforce compliance with the Consent Decree. Under the terms of this order, GBB took control of the Guam Solid Waste Management Division and assumed all of its responsibilities. GBB is presently managing the closure of Ordot and the construction of the new Layon Landfill.

The Layon landfill is scheduled to be completed and opened in the summer of 2011. In January of 2011, Senator Tom Ada introduced a bill to create an autonomous agency—the Guam Solid Waste Management Authority—to take over solid waste management responsibilities from GBB when the terms of the Consent Decree are fulfilled.

The total cost of constructing Layon Landfill and related consent decree projects was estimated to be $98 million in the August 2010 Quarterly Receiver Report.[39] GBB developed the financing plan for the process and assisted the Government of Guam with raising the funds. GBB is continuing to identify cost-saving measures for construction of the landfill.

To cover debt repayment and ongoing operations and maintenance costs, the new Solid Waste Management Authority will need to charge a tipping fee and maintain a minimum flow of municipal solid waste (MSW) into the landfill. In July 2009, the U.S. Navy, GBB, and the Government of Guam signed a letter of intent stating that it is the intent of the U.S. Navy to negotiate a contractual arrangement for solid waste management services with the new entity that will run the landfill. Essentially, U.S. Navy waste is needed to help meet project cash flow and revenue requirements for landfill operations and bond repayment. If U.S. Navy waste is included, it will also lower rates for the rest of the island. Final tipping fees have not yet been approved by the PUC[40]. However, proposed rates have been submitted to the PUC and are shown in

The impacts of increased tipping fees on waste collection rates are not known at this time. However, it is likely that higher tip fees will lead to increased illegal dumping and reduce the amount of waste going to landfills. Depending upon the cost of recycling, it may also lead to increased participation in recycling programs.

Table 7. Proposed and Interim Tip Fees

Landfill Tip Fees:		
	With Military	Without Military
Commercial ($/ton)	$ 156	$ 174
Residential ($/month)	$ 28	$ 32
Transfer Station Rates:		
Effective Date	Self Drop (< 3 CY)	Self Drop (> 3 CY)
Current	$ 2.50	$ 5.00
September 1, 2010	$ 4.00	$ 8.00
March 1, 2011	$ 5.50	$ 11.00
September 1, 2011	$ 7.50	$ 15.00

2.2.5.1. Guam Waste Generation and Composition Analysis

Minimal data on the amount and format of Guam's solid waste stream is available. In the August 2010 Quarterly Report, the Receiver reported that from June 1, 2009, through June 1, 2010, 80,500 tons of solid waste was delivered to Ordot. This is quite a bit lower than a previous estimate. In a report entitled *Estimation of Potential Landfill Gas Yields for the Ordot Dump*[41] , Duenas Associates and URS Corporation estimated that between 2004 and 2007 approximately 120,000 tons per year were being (or would be) disposed of in the Ordot Dump. As the Duenas/URS figure was based on estimates and the 2010 figure is based on scaled data, NREL will assume an annual waste estimate of 80,500 tons/year. This figure does not include green waste and corrugated cardboard, which were banned from Ordot several years ago. These materials are now being recycled by the private sector on Guam.

This annual total also does not include DOD solid waste. The amount of waste generated by DOD is estimated in the Final Environmental Impact Statement for the build-up[42]. The quantity of waste produced by DOD is highly dependent upon the population of military and non-military personnel associated with the build-up and related activities. DOD is also implementing an aggressive recycling program, with a target diversion rate of 50% by 2015 and every year thereafter. It is presumed that some of the material diverted to recycling would also be potentially suited for conversion in a waste–to-energy (WTE) plant.

The table shows a total solid waste volume from DOD and non-DOD personnel and construction activity starting at 30,467 tons/year in 2010 (including 10% recycling of DOD waste) and reaching a peak of 66,900 tons/year in 2015 (including 50% recycling), and then declining to 37,425 tons/year in 2019. For purposes of this report, NREL will assume the lowest value, or 30,467 tons/year of waste produced by DOD. This value is potentially conservative and there may be significantly more material available from DOD for a WTE plant.

Thus the total potential waste stream available is 110,967 tons/year. This is the number that we will use later to estimate the potential energy that could be produced from Guam's waste.

To our knowledge there are no recent, statistically-valid waste composition analysis studies that have been conducted on Guam's waste stream. In its December 2010 Quarterly Report, the Receiver provides information on a new waste characterization study being performed by the Solid Waste Management District (SWMD)[43] . This is a limited study in that it is only focusing on the waste generated by participants in SWMD's pilot recycling program. As part of the pilot recycling program, some residents have agreed to have their waste collected separately and sorted into categories and weighed over the next year. The first sort was conducted on October 12, 2010. A total quantity of 457 pounds of waste from eight residents was sorted into categories. The breakdown of materials into categories is shown in Figure 17 . Glass and metals (non-combustible) represent 13% of the stream.

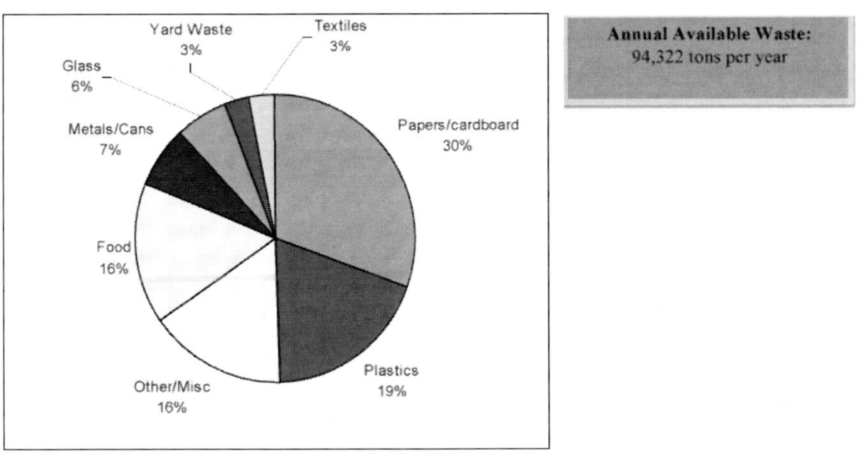

Source: NREL.

Figure 17. October 2010 waste characterization study (limited sample).

In addition, NREL located a reference to a 1995 study quoted in a memo from the Receiver; the Receiver provided NREL with a copy of this report[44] . The 1995 report estimates that Guam produces over 600 tons of solid waste per day. However, this number is not based on scaled data and it is considerably higher than the actual scaled data reported by GBB for the 2009-2010 timeframe. The 1995 study did conduct a detailed waste sort and characterization. The study found that approximately 13% of the material was either metal, glass, aluminum, or hazardous material.

NREL also located a 1982 study that categorized Guam's waste in 1980. The breakdown of the 1980 data is shown in Figure 18. Comparing the limited sample from 2010 with the analyses conducted in 1980 and 1995, glass and metals (which are non-combustible and should be removed from any calculations estimating available WTE conversion) were estimated at 13% of the total in 2010 and 1995, and 15% of the total in 1980.

As the 2010 sample is only from eight households, and pending the location of any more recent data, we will use the 15% value from the 1980 study as being the most representative and most conservative.

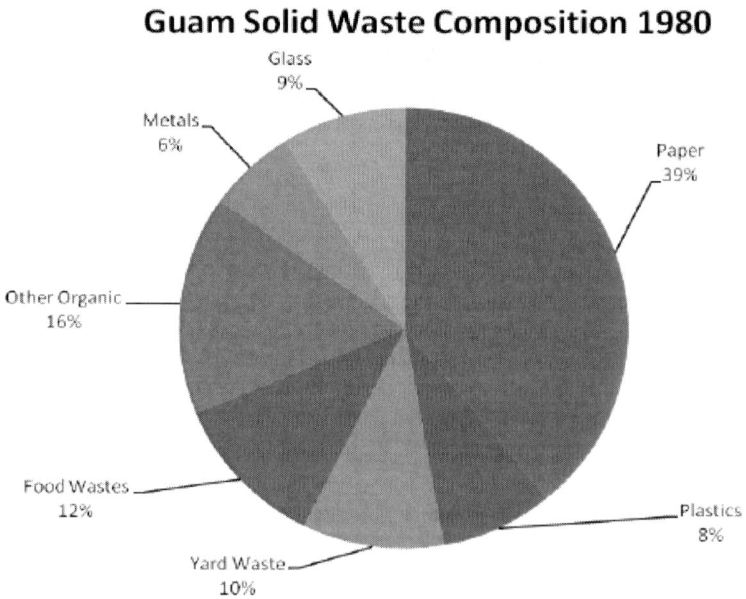

Source: Rossi Nayve Consulting Service and Associates.

Figure 18. Guam's solid waste composition, 1980.[45]

Assuming that the composition is still the same, and that the percentages also apply to the DOD stream, about 15% of the stream is non-combustible metals and glass and should be subtracted from the 110,967 tons per year of available waste. This leaves a potential available feedstock of 94,322 tons per year. We will use this value to estimate WTE potential in Section 3 of this report.

2.2.5.2. Construction Waste

There will be considerable construction and demolition (C&D) waste associated with the military buildup on Guam. Section 18.2.5.1 of the EIS document states that the preferred alternative will be to dispose of this material in the U.S. Navy-owned landfill at Apra Harbor[46] . While there will be a temporary increase in C&D waste associated with the build-up, the quantity will vary and it is not a reliable feedstock that will be available over the 20-year life of any WTE plant, so it will not be added to the estimate of available waste for conversion to energy.

2.2.5.3. Recycling

Recycling provides a landfill diversion opportunity, as well as an economic opportunity within the commercial sector. Observations on-island and interviews with stakeholders indicated there is presently minimal recycling in Guam. Should a more comprehensive recycling program be implemented, the amount of material available for WTE could potentially be reduced depending on the materials that are recycled. Solid waste managers and planners need to carefully consider all of the costs and benefits of both WTE and recycling within a unified framework. The lifecycle economics of recycling on Guam need to be carefully evaluated. On-island markets are likely to be limited and the costs of shipping overseas need to be considered.

2.2.5.4. Regulatory Framework

During the site visit, the legalities surrounding PL 25-175 and the ramifications were noted. PL 25-175, "An Act to Approve and Amend the Integrated Solid Waste Management Plan," prohibits construction or operation of solid waste incinerators or WTE facilities on Guam. Specifically, PL 25-175 states:

> Section 73113 is hereby added to Chapter 73, Division 3 of Title 10 of the Guam Code Annotated to read as follows:
>
> Section 73113. Municipal Solid Waste Incinerators Prohibited. Notwithstanding any other provision of law, it shall be unlawful for any person to construct or operate on Guam a municipal solid waste incinerator or a waste-to-energy facility, as defined by the rules and regulations of the

United States Environmental Protection Agency or the laws of the United States of America[47].

NREL understands that the main reason this legislation was passed was to prohibit implementation of a controversial 1996 agreement between the Government of Guam and a company called Guam Resource Recovery Partners (GRRP). The agreement was entered into by the Executive Branch of the Government of Guam. It is NREL's understanding that the legislature had concerns that the agreement required the Government of Guam to provide GRRP with minimum quantities of waste that were beyond the amount of waste generated on the island. The legislature therefore passed a law prohibiting WTE.

It is clear that without legislative action or creative thinking, it is extremely unlikely that anyone will be able to develop a WTE plant on Guam in the near term. However, we will still estimate the potential energy that could be produced from waste in Section 3.

3. Energy Efficiency and Renewable Energy Opportunities

Guam represents a high priority location for the development of an integrated energy plan due to its strategic importance to the U.S. military, high energy costs, dependence on imported fossil fuel, and the expected future pressure on the island's energy resources from the planned increase in military and civilian populations.

There are many energy efficiency and renewable energy opportunities for Guam. This section outlines the opportunities that have been identified based on the initial assessment. However the Task Force will need to determine where priorities lie and which opportunities will need to be considered further. The initial opportunities identified are summarized in Table 1. Energy Efficiency and Renewable Energy Opportunities and Potential Impacts.

This section of the report provides information on various opportunities as they relate to developing a strategic plan. Specific savings are difficult to quantify due to the amount of data available at the time of the site visit. Savings could be calculated based on energy audits from private industry and scaled up across the sector, or more detailed audits could be conducted to determine specific savings. The areas of opportunities presented in this section are based solely upon observations made during the site visit conducted by the NREL team while on-island the week of October 11, 2010.

The energy efficiency and renewable energy opportunities discussed below are:

- Energy efficiency and conservation
- Wind
- Solar
- Geothermal
- Biomass.

3.1. Energy Efficiency and Conservation

The opportunities that have been identified for energy efficiency and conservation measures include raising island energy awareness in the various sectors, creating a cool-roof program, and establishing metrics for improvement within the island building codes and equipment standards.

Due to administration changes, it is unclear at this time the direction Guam will take on building codes. It is recommended that a review of the existing building codes and equipment specification standards be conducted by the Energy Task Force, utilizing available technical assistance. The end goal of a building code and specification analysis would be to highlight areas where improvements could be made and where Guam could benefit by changing the building code and standards. Other island nations with energy efficiency standards already incorporated into codes and standards could be used as a model. Energy efficiency integrated into tropical building codes, training, and enforcement are all potential next steps after a review of the existing codes is conducted.

3.1.1. Raise Island Energy Awareness for Conservation

Energy audits have been funded on island, most recently, by the Guam Energy Office (e.g. audits of government facilities and homes), the Department of Education (schools), and by GPA and Johnson Controls (e.g. hotels, the airport, etc).

It is recommended that building audit results from the current activities on Guam (e.g. within hotels, schools, government agencies, and homes) be applied to developing a strategy where conservation measures are relevant to multiple buildings within a specific sector. For example, hotel audits may indicate the need for better lighting controls; this should be integrated into a strategy to improve controls in all hotels, not just those being audited.

These audits may identify both behavioral and technical improvement and are summarized by sector below.

- Residential
  - Campaigns for local communities, as discussed in more detail in Appendix B
  - Pilot the campaign in Dededo where there's interest and the potential for the largest impact as discussed in more detail in Appendix B
- Commercial
  - Work with various sectors to increase awareness through posters, stickers, fact-sheets, etc.
  - Work with island suppliers and retailers to source energy efficient technologies
- Construction industry
  - Provide training for the construction industry on design practices and energy efficiency measures that should be incorporated into new construction and renovations, such as external shading for windows, coating coastal air conditioning unit coils to protect from sea mist, insulation, high SEER air conditioning units, etc.

3.1.2. Establish a Cool-Roof Program to Improve Efficiency

DOE's Secretary Steven Chu launched a series of initiatives in July 2010 to encourage cool-roof technologies in the federal government. Secretary Chu has directed all DOE offices to install cool roofs whenever cost effective over the lifetime of the roof, when constructing new or replacing old roofs, and in areas where cool roofs will be beneficial to energy efficiency savings (typically in hot climates).

While these initiatives are targeted at federal buildings, the cool-roof program has materials and resources that would be beneficial to Guam. Cool roofs are roofs that are designed to maintain a lower roof temperature than traditional roofs while the sun is shining, typically through reflectance technologies. Not all cool roofs need to be white, as indicated by the cool-roof tiles in Figure 19[48].

A cool-roof program could be implemented island-wide by using low-cost technologies, such as Elastomeric paint (~$40/gallon), which would need to be applied to roofs and maintained through a cleaning program (to prevent debris and mold build-up). More information about selecting buildings and implementing a cool-roof program can be found on the DOE website[49].

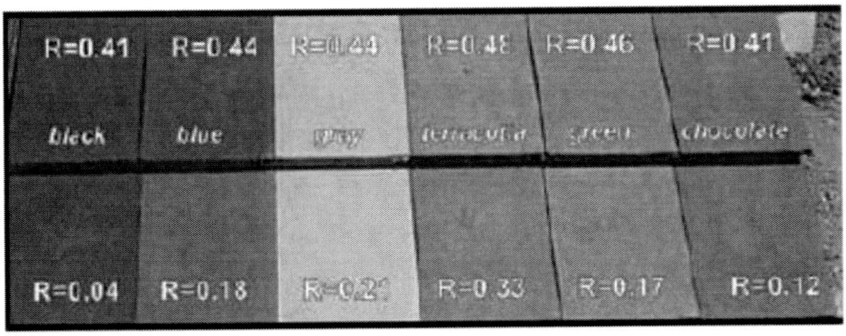

Cool-colored tiles (top row) look just like conventionally colored tiles but have higher solar reflectance (R). *Image Source: American Rooftile Coatings and Lawrence Berkeley National Laboratory*

Source: DOE.

Figure 19. Cool roof tiles in colors other than white.

3.1.3. Review and Improve Energy Efficiency Standards to Improve Efficiency

Building codes were accepted by government officials of Guam; however, energy efficiency should be incorporated into building codes. Building codes should be reviewed and revised to include energy efficiency parameters as soon as possible. The building industry will need time to learn building codes and train personnel to meet these codes, so integrating energy efficiency as early as possible will assist with a faster implementation rate of energy efficiency measures in new construction. Equipment standards should also be reviewed to eliminate/reduce energy intensive equipment, such as window air conditioning units and machinery being imported to Guam. The European EcoLabel scheme or ENERGY STAR ratings are good models to follow for establishing equipment and buildings that meet specific criteria for meeting Guam's energy reduction targets. The EPA Portfolio Manager is useful for managing building stock and benchmarking buildings across a portfolio.

3.1.4. Consider Alternative Cooling Technologies

Cooling is one of the highest end users on Guam within all sectors. Considering and implementing promising alternative cooling technologies will assist with reducing fossil fuel consumption related to conditioning indoor spaces. There are a number of options available under the umbrella of alternative cooling technologies.

One lower-cost option in buildings with no requirements for conditioning for moisture control is to install ceiling fans. Where moisture buildup is not a concern it may be possible to utilize natural ventilation (from breezes and open structures of windows, doors, etc.), and ceiling fans, which are in operation only when needed. There may be open structures where this could be a suitable option.

Planting 'green roofs' may also be a lower cost option to consider. Where roofs are flat and structurally sound they can be used to plant vegetation, which not only provides a thermal barrier, but also creates a wicking-type action. The wicking of moisture and heat upwards and out of the building is beneficial in reducing cooling loads in buildings[50].

A medium-cost option is to install solar air conditioning units. While relatively new on the market, solar air conditioning units are available from a number of manufacturers and suppliers, and have a variety of different installation applications from split wall window units to larger outdoor units. This technology is an efficient type of air conditioner (upwards from a SEER 20) that operates off of a solar photovoltaic (PV) panel, which reduces the energy demand from the grid during times when cooling and conditioning are needed.

A high-cost option to reduce Guam's energy consumption related to space cooling is a sea-water cooling system. This technology is being investigated by private consultants for the U.S. Navy and also by GPA and JCI for the hotel loop in Tumon Bay. The Energy Task Force is encouraged to review the findings of that investigation to determine whether the cost-benefit analysis is in the island's favor.

3.1.5. Next Steps

The following next steps are recommended:

1. Continue energy and water audits and apply measures where they are more broadly applicable.

Audits have already begun on Guam in specific sectors, such as the airport, Joint Region Marianas sites, and within a few hotels. These audits should continue and, where possible, the measures that are applicable to similar buildings should be gleaned from audit reports and implemented across the building stock. For example, if five of the island's hotels are being audited the measures that are most commonly recommended may be applicable to the remaining hotels. These measures may include improving lighting fixtures and controls; installing low-flow faucets, toilets, and showerheads; or installing thermostatic controls to reduce the energy consumption of air conditioning units. A strategy for identifying and implementing the measures that are most commonly recommended could be a next step, but would require disclosure from the entities conducting the energy and water audits.

2. Begin an energy awareness campaign within the various building sector types.

An energy awareness campaign is a low-cost solution for implementing energy conservation. Pilot projects could be carried out within specific buildings or communities to highlight the impact of the campaign on energy consumption. For example, Dededo has an actively engaged Mayor who may be willing to promote energy conservation. In order to begin an awareness campaign the audience would need to be identified, materials prepared, and launch dates selected. Campaigns are ongoing and need to be renewed every few months to communicate specific messages to target audiences. Materials within a community such as Dededo may include articles in newsletters or the weekly newspaper insert, posters and flyers, stickers for light switches and equipment, and events for children and adults that are fun and educational. A different campaign should be created for each specific audience and may include the following:

A. Hotel staff and guests
B. Medical staff
C. Education staff, teachers, and students
D. Community residents (adults and children).

3. A strategy for implementing a cool roof program should be designed, beginning with identifying buildings to be used as pilots for a cool roof pilot program.

These buildings should be selected following the guidelines established in the materials on DOE's website[51]. Buildings would ideally have engaged facilities management staff who could monitor the energy consumption prior to and after the cool roof installation to indicate any change in energy consumption. Buildings that vary with sector and cool roof application type would be most beneficial to display what is most effective at reducing energy consumption. Facilities in need of roof replacement, particularly if there is a budget for replacement in the near-term, may be most applicable as initial projects. Once this pilot program has been completed the results should be analyzed and the most effective cool roof type should be replicated throughout the building stock, with continued monitoring and savings reports.

4. Create training materials for the design and construction industry to increase knowledge relating to energy-efficient technologies and practices.

Training materials could take the form of curriculum for workshops, handbooks, presentations, websites, and other tools. A training module would need to begin with an assessment of the current level of knowledge and indicate areas of the greatest need for improvement. The audience would need to be assessed so that materials could be tailored to be the most effective.

5. Encourage the use of alternative cooling technologies, where appropriate, within retrofits and new construction.

Review the results of the salt-water cooling research and determine if implementation in specific industries (e.g. hotels) is beneficial to reducing fossil fuel consumption.

6. Develop a strategic energy plan.

There is great opportunity for policy and technology development in Guam. Through the assistance of Guam's Energy Task Force and the development of a strategic energy plan, policy and technology can be addressed. Ongoing efforts will be needed to support the Task Force on a multitude of levels. Although the Task Force has been established, Guam has currently elected a new governor and administrators; this change can set timelines back. However, Governor Calvo and his administration were quick to be informed and fully support this effort. Maintaining momentum is key to success. The Task Force will require some assistance in deciphering data and directing subcommittee levels while setting goals and objectives. Several deeper dives within technology development and cross-cutting policies will need to be examined to determine best strategies.

7. Develop a policy framework strategy.

Policy is an effective tool in addressing clean energy adoption and subsequent development, and the process of policy development addresses the interconnectedness of social concern. Below is an outline of a policy framework strategy Guam may want to use in developing its strategic plan.

- How is the energy discussion being framed?
    - What high-level support is there (federal, territorial, local)?
    - Identify the support; finances, technical assistance, education and outreach activities.
    - How does clean energy address island core values? Find and identify all the connections
    - Is there energy literacy in the community, in the business sector, and in the local trades?
- What are the energy goals (federal, territorial, local)?
    - This should cover all sectors; water, agriculture, aquaculture, recreation, tourism
- Stakeholder analysis
- Identifying all barriers
    - Social awareness and acceptance
    - Financial
    - Technical
    - Workforce
    - Education
- Surveying current policy
    - What are Guam's policy nuances?
    - Are they effective?
    - Is there a local-territorial government connection?
- Policy analysis
    - What technologies support Guam's goals, objectives, and budget?
    - What are the policies and programs to support those technologies?
    - What existing policies support or hinder Guam's energy goals?
- Implementation and action plan
    - Implementation strategies
    - Budgets and timelines
    - Measurement and verification
    - Re-assessment

Addressing all these topics through the development of a strategic plan with the appropriate expertise sets a sturdy foundation for assessing opportunities and achieving energy goals.

3.2. Wind

Wind energy has the potential to play a major role in supplying electrical energy to Guam. The available wind resource combined with a high level of infrastructure, high energy costs, and a strong federal presence on the island, make wind technology a very viable option. Additionally, several wind assessments have been conducted by both the U.S. Navy and GPA.

Through an Interagency Agreement between the Naval Facilities Engineering Service Center and NREL a 60 m met tower has been installed on the western boundary of the Ordnance Annex just below Mount Alifan. This measurement station is installed on an island high point andprovides an excellent wind resource reference for the island, the only such reference currently available. The measurement site was installed in September 2008 and thus far shows a very good class 5 resource on the island with a mean wind speed in the region of 7.9 m/s, with stronger winds during the winter months. A summary of the data collected to date is provided in Appendix C and clearly demonstrates that the island has a viable resource if an appropriate development site can be identified.

Although not verified, viable wind sites have been identified by GPA for up to 20 MW of installed capacity in a large-scale wind development, with the potential of several smaller, 1 MW to 5 MW scale projects at several locations around the island. Additional large-scale wind development project locations may still be identifiable, primarily on the northeast coast or buttressing eastern ridges on the southern end of the island. However, the identification of the GPA Cotal site, combined with the potential sites on U.S. Navy property provides a key focal area for near-term project development.

In a study conducted by DNV-Global Energy Concepts for GPA, the Cotal site area, which is located just to the north of the Cross Island Road (HWY 17) east of Mount Alifan and the community of Apra Heights. This site could support approximately 10 large wind turbines along two ridges, with a total capacity of approximately 20 MW. The site has good road access, is close to a 34.5 kV transmission line, and is within several miles of a substation. The site was reviewed during the NREL island visit and proved to be a good location. The turbines would be installed upwind from several homes and a larger potential development area. The site sits directly south of the Leo Palace Resort and although not close, will be clearly visible from the resort. The elevation of the site is around 200 m, which puts it only approximately 50 m lower that the island high spots and approximately 3 mi. from the U.S. Navy met mast, which shows an annual wind speed in the area of 8 m/s. Land in this area is primarily owned by the Department of Agriculture with several smaller private holdings.

As stated, several other sites have been investigated though various assessments, including on the south end of the island, along the main island ridge near the U.S. Navy magazine, on the Apra Harbor breakwater, land to the south of the Cross Island Road (HWY 17) in the area of the Cotal site, and some locations along the northern and northeastern coast. Most of these sites could be given additional consideration although all of them have one or two critical siting shortcomings, such as a reduced or more turbulent wind resource, limited local transmission infrastructure, or more complex siting, indicating that none of the sites identified thus far seem to be as viable as the Cotal site.

More information regarding the wind assessments that have been conducted, including a summary of data collected through the U.S. Navy-sponsored project and the GPA-funded studies is provided in Appendix C.

3.2.1. Key Challenges to Wind Development

There are several key elements that need to be taken into account when considering the deployment of wind technologies. In the context of Guam, environmental impact, specifically relating to threatened species and ground cover, radar/communications impacts, social acceptance, and the impact of wind on grid stability will be critical components for wind deployment on the island.

The environmental and radar impacts of wind deployment are two of the most critical due to their binary nature in that if the proposed site and/or turbine is likely to have adverse environmental or radar impact, the site is likely non-viable. Grid impacts are primarily technical in nature and can be addressed using a number of techniques, many of which will have a cost impact on the project. Due to the high cost of conventional power generation it is likely that even with the additional cost of power control system modifications, the wind project will remain financially viable, reducing the likely impact of this barrier.

Social acceptance of wind technology could become a major issue in successful project development and thus must be carefully orchestrated. Due to the small size of Guam and generally high population density, any development site, primarily in the north, will result in a social impact. This impact will likely be in the form of noise and visual clutter; although infrasound, adverse health, and flicker (the strobe effect of shadow caused by the wind turbine blades) could also be brought up as concern. Local opposition could develop around these issues if people felt that the site of the wind turbines would negatively impact island life. This is especially true if the owners of the Leo Palace Resort felt that a proposed project at Cotal would impact their business model, as this could represent a funding mechanism for a local anti-wind development advocacy group.

Generally speaking, project implementation is greatly improved if the local community feels that the project provides community or personal benefit. The fact that the wind project can represent a means to reduce the cost of power to all of the island's communities and that it would reduce dependence of imported energy sources will decrease the likelihood of potential community criticism; however the control of this communication will be critical for eventual project success. Examples of wind developments on other islands, and input from residents of those communities regarding perceived and actual impacts, can assist in addressing potential societal concerns.

3.2.2. Infrastructure

The infrastructure for the development of large scale wind deployment on Guam, although problematic, is not seen as a critical issue. Although this may differ for other island locations, in the context of a potential project at Cotal, the road, transmission, general construction, and shipping infrastructure on-island is adequate for large wind turbine implementation.

The only current limiting factor is the availability of a heavy lift crane to support turbine installation. It could be that such a device will be available on-island due to the pending military-driven infrastructure improvements, or a crane will have to be shipped to the island. In either case this issue will reflect on project economics but will not impact the technical viability of the project.

Typhoons do pose an important risk in the use of wind technologies. All larger wind turbines are designed to stop operating in high wind conditions, meaning that in all except the most extreme cases storm-related damage is going to focus on that caused by airborne debris, typically to the turbine blades. The other potential issue would be the wind loading on the turbine structure exceeding the tower or foundation design loads, which are controllable factors, although they can cause increased implementation costs representing an engineering design issue. Although there has not been much experience with large turbines in the Pacific, a number of turbine installations in the Caribbean have not reported major issues, including a small system of four turbines installed at Guantanamo Bay on the eastern end of Cuba, an area with common hurricane (typhoon) activity.

Looking at the implementation of turbines installed in areas with known typhoon or extreme weather events, consideration must be given to both turbine selection and an assessment of the risk for the turbine structure, which may lead to the purchase of insurance. Every large wind turbine is designed around an international standard for wind intensity, provided in a class rating. The class rating will also impact the size of the wind turbine rotor, which will have an impact on the resulting energy generation from the turbine.

At least two manufactures have a typhoon-rated wind turbine, although it is not clear if this relates to Class V typhoons. Additionally the wind turbine manufacture Vergnet makes a 220 kW turbine that is designed to be lowered during potential typhoon events and is production testing a larger turbine where the rotor of the turbine is removed. In all cases however, the purchase of insurance, or in cases of self-insured companies, an assessment or risk is used to assess the appropriateness of installing wind in any environment.

Historically insurance companies have not been willing to provide insurance for turbines in Class V typhoon areas, but with the development of turbines rated for typhoon environments, it may now be possible to do so. In any case, one of the first steps in the development of potential projects will be to contact wind turbine manufactures, either informally or through a competitive process, to better determine what products are available; survival wind speed, as well as typhoon experience, should clearly be one of the key competitive requirements.

3.2.3. Next Steps

The following next steps are recommended:

1. Conduct a detailed wind development assessment for the Cotal site.

With the expectation that the development of this site will be conducted through a competitive request for tender, the more information that is available about the site will result in improved project design and economics. This activity would lay the groundwork by conducting all of the initial screening studies that will be required for site development. Other elements of the study would include:

- Resource assessment combined with data correlation. This would include the installation of a 50 m or 60 m met mast for a minimum of one year. It will also be critical to an overall analysis to correlate the wind resource data with the U.S. Navy Mt. Alifan tower. If at all possible, both met towers should be installed for a full overlapping year.
- Land title assessments
- Initial environmental impact assessments
- Geotechnical assessments
- Initial grid interconnection assessment
- Initial social acceptance documentation (noise and visual simulations)
- Typhoon risk analysis and impact of economics of wind generation.

This activity would be followed by more detailed feasibility and economic assessment of the project. Aside from the resource assessment, these activities should cost approximately $50,000. Additionally, costs of resource assessments can be significantly reduced if the met tower is classified as a temporary structure and not subjected to typhoon building codes.

2. Support a direct collaboration between GPA and the U.S. Navy regarding wind project development on both military and non-military property.

The military and GPA have overlapping activities that should be coordinated to spur collaboration. Additionally, the installation of wind turbines will have some impact on military operations, so these issues should be addressed at the onset of the project.

3. Initiate an active social acceptance outreach and education activity around wind.

For successful project development it will be critical to get ahead of the information around the deployment of wind technologies on Guam. This would include the quantifying of potential impacts, such as job creation, economic development, and consumer savings while developing information regarding potential negative impacts, such as conducting a noise impact assessment and visual simulations. All of this information should be publically available with open forum discussions conducted by GPA or another organization to identify and address potential issues. If at all possible this should be implemented during the initial assessment phase (within the next year) to get ahead of the development of inaccurate preconceived notions. This activity and the information generated through it should not directly advocate for a wind project, but should provide fair and balanced information on the impacts of wind development.

4. Expand investigation of additional potential sites for community and large wind development.

Given the potential savings that could result from expanded wind deployment on the island and clear access to a viable wind resource, efforts should be made to identify additional potential sites on Guam for expanded wind development in addition to expanded collaboration on sites being considered by the U.S. Navy. This would include an island-wide screening study using land ownership, resource through an island-wide wind map, environmental sensitivities, and known development limitations. This would allow the potential identification of additional locations that could be developed at varying degrees of cost. This information would then be used to support longer range strategic planning.

5. Conduct an initial screening of potential turbine options for typhoon environments.

If wind development is to be seriously considered, this would include both an assessment of available equipment as well as a risk assessment considering the frequency of typhoons of different classes. Through turbine manufactures or independently, insurance companies could be contacted to determine if insurance policies for wind turbine projects in Guam could be obtained and, if so, at what cost.

3.3. Solar

Solar energy technologies are often broken down by solar thermal and solar PV applications. Solar thermal technologies capture energy from the sun to heat a fluid or air, which is generally used to heat water. Solar PV technologies are used to capture the photons from the sun to create electricity. These two technologies and their applications are outlined in this section.

3.3.1. Solar Thermal

Solar thermal technologies, also known as solar hot water, are typically low- to medium-cost and easy to install. Conversations with several hotel engineers on Guam have indicated that the second largest energy cost for hotels is hot water (after air conditioning and chiller loads). The current practice on Guam is to heat water using diesel fuel, although some hotels on Guam are using vegetable oil to heat water. Solar hot water is a low to medium cost application that should be considered within the larger industries where hot water consumption is high.

Figure 20 shows a common solar thermal system used for hot water heating[52].

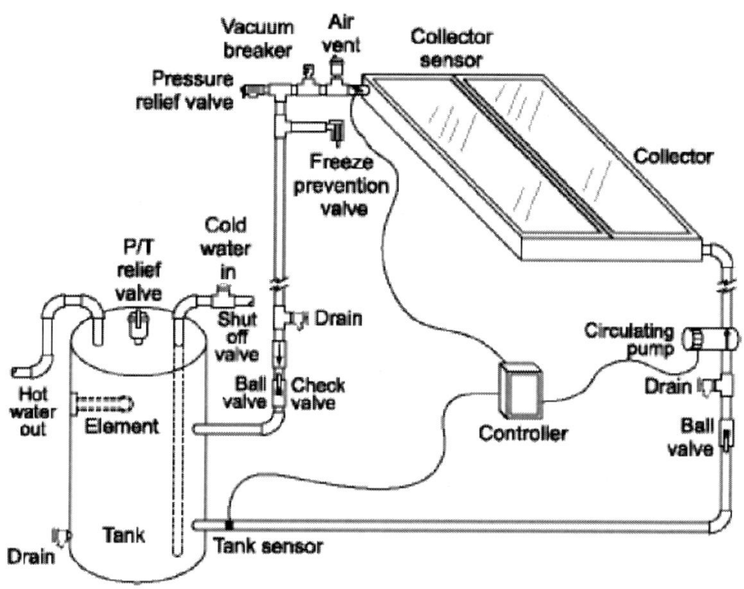

Source: NREL.

Figure 20. A common solar thermal system.

A solar collector captures the heat from the sun and then gravity typically draws a fluid out of the collector to act as a heat exchanger to transfer the heat to another material, such as water in a tank. Pressure relief valves and circulation pumps are needed to maintain pressure and keep the fluid moving through the distribution system. Basically, these systems convert sunlight to thermal energy by absorbing the sun's radiation to produce heat which is transferred to water. Like most technologies, there are several types of systems that follow this basic operation and can be applied to various applications in both residential and commercial sectors. This technology works well in recreational settings such as pool heating.

Solar thermal heating works well in most situations. The system efficiency is improved with the use of evacuated tubes, however even low-tech collectors are able to generate a low grade heat, which offsets the amount of energy required to heat water in most applications. By reducing the amount of hot water being generated through conventional forms of heating, the demands on fossil fuels can be lessened.

Solar thermal technologies should be considered for hotels, hospitals, laundromats, assisted living facilities, and residences where hot water demands offset the capital costs of installing solar thermal panels.

Panels should be installed where there is exposure to the sun without shading from buildings or vegetation. They should also be installed in locations where they will not be vandalized or tampered with, and they should be installed according to building regulations, plumbing, and access codes[53] .

Source: NREL/PIX 09188.

Figure 21. Solar hot water, Hickam Air Force Base.

As a comparitive case study, the United States Coast Guard in Honolulu, Hawaii, installed 62 active/direct solar thermal system units in 1998. The average cost at that time was $4,000 per system with an additional discount of $800.00 from the utility. The utility had previously initiated a rebate program as an incentive for solar hot water installation for residences. The energy savings was 3,008 kWh per year with a demand savings of 1.62 kW per house.[54] Figure 22 refers to the average hourly power per household within a 24 hour period.

With the current market, solar thermal systems have reduced significantly in price and increased in efficiency. With the end-use statistics mentioned in Section 2.2.3, both commercial and residential solar hot water systems are opportunities for fossil fuel replacement and energy savings.

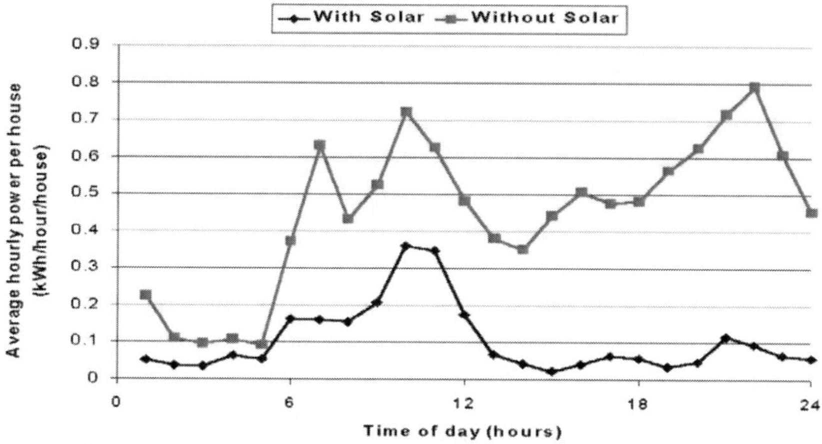

Source: NREL.

Figure 22. U.S. Coast Guard housing, Honolulu, Hawaii.

3.3.2. Solar PV

Winzler and Kelly, on behalf of GPA and in support of GPA's Integrated Resource Plan (IRP), evaluated various Guam properties for potential use as a PV solar power plant (see Figure 23). This evaluation will be included as an attachment to GPA's upcoming Invitation for Multi-Step Bid (IFB) for renewable energy resources to provide guidance and background information to potential solar power plant developers. Solar power is just one of several options that will be considered under the IFB.

The criteria that were used to evaluate sites were developed in conjunction with GPA and R.W. Beck, which is also providing consultation services in support of GPA's IRP. It should be noted that this evaluation is only presented as a guide and does not establish any specific requirements that potential solar power plant developers must follow to satisfy the requirements of the IFB. The criteria and resulting evaluation of sites contained herein are merely presented as a starting point for potential developers to utilize as they desire. It is also important to note that no land owners of sites evaluated herein have been contacted nor has any effort been made to evaluate potential real estate or other costs.

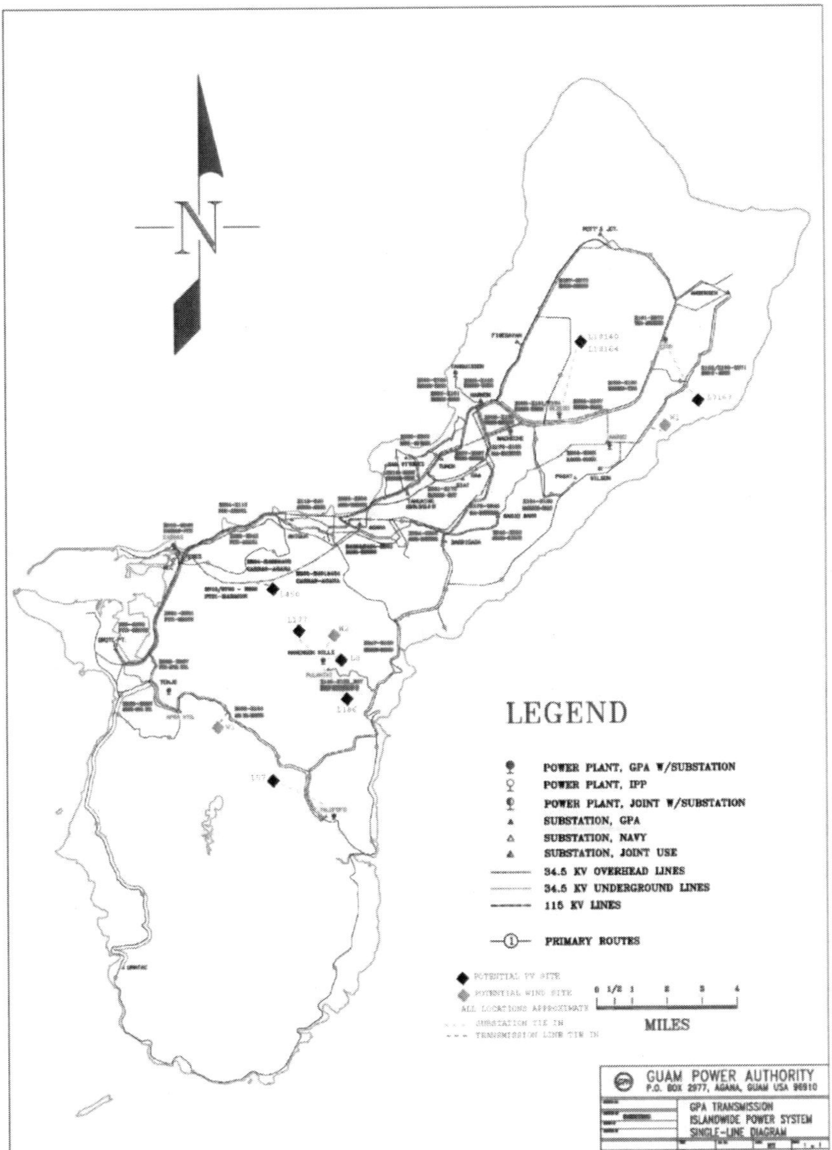

Source: GPA.

Figure 23. CPA transmission system map with potential solar and wind locations marked.

In general, the scope of work for this evaluation included the following:

- Investigate between three and six sites outside of DOD property. Note: R.W. Beck provided footprint information of 60-80 acres per 10 MW plant, to which GPA gives preference
- Review topological data, access roads, proximity to GPA transmission/distribution systems, other improvements, and land use and property ownership (based on publicly available information at the Department of Land Management, no title reports required)
- Submit Technical Memorandum that describes advantages and disadvantages for the potential solar development locations, including potential plant sizes in investigated areas.

3.3.3. Next Steps

The following next steps are recommended:

1. Create a subcommittee within the Energy Task Force to address all possibilities of policies and programs for solar hot water systems for all sectors.
2. Consider a renewable energy interconnection study.
3. Conduct field assessments of each site to determine how many acres of each site can actually accept solar arrays, and to determine the potential energy output (in MWh) from those more detailed solar plants assessments.

Table 8. Top Seven Sites Identified for Solar PV

Lot Number(s)	Total Size	100-kW / acre	200-kW / acre
	acres	kW PV	kW PV
L450-R3	1,125	112,450	224,900
L177-4-R2	1,072	107,200	214,400
L7163	416	41,550	83,100
L10140-R5 and L10164-R3	191	19,070	38,140
L57-2	169	16,890	33,780
L0	111	11,090	22,180
L186NEW-1	72	7,150	14,300
Totals	3,154	315,400	630,800

The top seven parcels for installing solar arrays identified during this study are shown in Table 8. The third column is an estimate of the total amount of the array output at a conservative (relatively low) value for the array density, and the fourth column is an estimate at an optimistic (relatively high) value. Assuming every acre of every lot could be populated with PV modules, the potential for solar PV could range from 315 MW to 631 MW.

3.4. Geothermal

Due to the enormous need for clean baseload generation and the increasing need for energy diversity, geothermal potential must be assessed and determined to be a likely technology to implement or not. Baseload geothermal power would be a tremendous asset if it could be found and developed on Guam.

A team from NREL and the U.S. Navy's Geothermal Program Office in China Lake, California, conducted a reconnaissance assessment of the geothermal potential of the island for DOD in April 2010.

Guam has no obvious surface features suggesting geothermal potential and has never been the focus of a geothermal exploration campaign, nor even a rigorous geothermal assessment. Guam lies on a regional trend of high heat flow, the Izu-Bonin-Mariana (IBM) volcanic arc system. However, while the central and northern portions of the IBM arc are volcanically active, Guam is not, and has not been, for millions of years. Nonetheless, active submarine volcanism occurs in deep water tens of miles west of Guam.

The NREL/U.S. Navy team uncovered a reliable report of a steam vent in an abandoned limestone quarry on the southwest corner of Tumon Bay in fall 2010. The team located the quarry and spoke extensively with Professor John Randall, University of Guam (retired) who visited the quarry in the early 1970s specifically to observe the steam phenomenon. The evidence of the steaming vent, and a report of a hot water well in the Ylig Valley, suggest that geothermal fluids are present in the subsurface on Guam. Certainly, there appears to be enough information to warrant further investigation and exploration for geothermal systems on Guam.

3.4.1. Next Steps

The U.S. Navy is conducting an assessment to identify potential geothermal locations using Light Detection And Randing (LIDAR) technology during spring 2011. The U.S. Navy has committed funds to drill temperature gradient holes starting in summer 2011. The specific locations where the temperature gradient holes will be drilled will be guided by the results of the on-going LIDAR study.

3.5. Biomass

Biomass is generally defined as any organic feedstock available on a renewable basis. Typical biomass resources include wood and wood waste, MSW, landfill gas (LFG), agricultural and crop residues, human solid waste, and animal manures. One of the primary benefits of biomass is that feedstocks can be stored, thus allowing for baseload, dispatchable generation. If using a biofuel, such as biodiesel, the fuel can be used directly in the existing equipment, thus allowing for renewable fuel switching at minimal capital cost.

Biomass resources that are potentially suitable for Guam include:

- MSW electricity generation Biodiesel or straight vegetable oil production and use in existing fuel oil boilers or engines
- LFG capture for energy recovery
- Anaerobic digestion to generating biogas at waste water treatment plants and small-scale digesters for converting animal manure to biogas
- Dedicated feedstock productionto produce solid or liquid biofuels

During the site visit, the NREL team evaluated potential opportunities to develop each of these options.

3.5.1. Waste-to-Energy

The generation and disposal of MSW presents challenges for most communities, but especially islands. Converting MSW to energy is one option that many communities have either undertaken or are considering. There are several challenges to developing a WTE facility on Guam, not the least of which is a law that was passed by the legislature in 2001, that effectively prohibits WTE projects in Guam. These challenges were outlined in Section 2.2.5 of this report.

As discussed in Section 2.2.5, our best estimate is that Guam produces approximately 111,000 tons of MSW per year, of which 94,321 tons are deemed to be potential feedstock. There may be other feedstocks available, such as tires, sludge from wastewater treatment plants, green waste, and some small quantities of agricultural residues. As these resources have not been quantified, and to do so is beyond the scope of this report, they are not included in the following analysis. There is also the potential for increased waste generation associated with the DOD build-up. These numbers must also be quantified for future efforts.

As shown in Table 9, we estimate that 94,321 tons per year of MSW would be sufficient to fuel a 4.7 MW to 7.0 MW (net output) plant with an 85% capacity factor. The actual generation will be highly dependent on the selected technology, the overall composition of the feedstock, how the feedstock is handled and prepared (e.g. is it processed into refuse derived fuel or used as is?), system design, and overall operating conditions.

Table 9. Summary of WTE Potential

Total Waste (tons/year)	110,967
Estimated non combustibles (%)	15%
Combustible waste (tons/year)	**94,321**
Capacity Factor	85%
Tons/operating day	304
Net power - low efficiency (MW)	6.5
Net power - high efficiency (MW)	9.7
Net energy - low efficiency (MWh)	48,103
Net energy -high efficiency (MWh)	84,889

The value for “low efficiency” in Table 9 is based on an assumed yield value of 510 kWh per ton of MSW processed. This is equivalent to a typical mass burn combustion plant today. The value for “high efficiency” assumes a conversion factor of 900 kWh/ton MSW. This is equivalent to what some manufacturers of advanced conversion technology gasifiers and plasma units are claiming to be able to obtain from their systems.

It should be pointed out that there are no commercial advanced conversion technology WTE projects operating in the United States, although there are several reported to be operating in Asia and Europe. Of the commercially-viable WTE systems that are operating in the United States, nearly all (if not all) of the plants use either stoker or fluidized bed combustion technology. Combustion technology is fully commercial and lenders are willing to make loans to well-structured projects that have a guaranteed supply of MSW and a power purchase agreement with a credit-worthy customer.

The numbers in Table 9 above assume that 15% of the output from the plant will be used for parasitic loads. In terms of net capacity, NREL estimates that between 6.5 MW and 9.7 MW of power could be provided from waste on Guam. Because WTE can be operated 24/7, a WTE plant produces more energy over a year than an equivalent wind or solar plant. So a 6.5 MW WTE plant would produce approximately the same amount of energy each year as an 18 MW wind farm (at 35% capacity factor), or 54 MW of solar at an 18% capacity factor. The WTE plant would take up far less space as well and could be co-located at the Layon Landfill.

The estimated capital cost of this project is $102 million. This is calculated based on assumed capital costs of $400,000 per ton per day. For comparative purposes, the 16.5 MW (net) WTE plant under development in the U.S. Virgin Islands is estimated to cost $160 million ($457,000/ton/day). This cost includes setting up two material recovery facilities for converting MSW to refuse derived fuel through a proprietary process called WastAway. Material from the two WastAway plants will then be sent to the centralized WTE plant. The developer states that the WastAway process creates a product that is dry and has a Btu content of 7,500 Btu/lb. The energy content of Guam's waste stream has not been reported but we estimate it to be similar to the U.S. average of 5,900 Btu/lb.

For Guam, NREL conducted a preliminary economic analysis of a low-efficiency WTE plant. Results indicate that if a private developer were to come in and build, own, and operate the plant, and the developer was not paid a tipping fee, electricity would need to be sold to GPA for $0.26/kWh (escalating at 3% per year) to make the project economically attractive. For the developer to be able to sell power to GPA at $0.15/kWh (with a 3% annual escalation), the developer would need to receive a tipping fee of approximately $75/ton.

It must be pointed out that the numbers above are preliminary and need to be investigated in greater detail. We are using rule-of-thumb numbers and there are many companies and technologies available. It may be possible for a developer to make the economics work using various technical and economic configurations. The best way to determine what the market can deliver is for GPA to develop a request for proporals (RFP) for WTE (after regulatory/legal issues are addressed) and see what bids are received.

The major barrier to WTE is PL 25-175. Until Guam changes the law and allows WTE projects to be developed, there is little to be gained through speculation on various technologies, configurations, technical performance, and cost. A more immediate short term need would be to focus on legislation that would allow flexibility on the development of WTE plants, bearing in mind local desires for clean technology and debt repayment on Layon Landfill. This could be an action for the legislative subcommittee of the Energy Task Force.

3.5.2. Biodiesel or Vegetable Oil

The use of biodiesel or vegetable oils in diesel generators and boilers is gaining wider acceptance around the world. The ability of biodiesel to be used within existing utility diesel infrastructure is a compelling reason for exploring this option. There are minimal capital costs associated with using a biodiesel or vegetable oil/diesel blend in existing diesel or residual fuel-oil-fired generators, combustion turbines, or boilers. Biodiesel blends of 2% up to 20% can typically be used in existing equipment with only minor modifications. There have been several testing programs, particularly in Hawaii, aimed at firing 100% biodiesel in utility-scale boilers, turbines, or generators. Several utilities have been running biodiesel in standby generators as a way to implement peak load reduction and reduce emissions associated with running the generators beyond their annual permitted hours.

The use of 100% biodiesel in utility-scale generators is being pioneered in Hawaii, where there are at least three different projects underway:

1. Hawaiian Electric Company (HECO) recently completed co-firing tests of various blend percentages of petroleum diesel and biodiesel fuels at the 90 MW Kahe Power Plant. On January 28, 2011, HECO issued a press release stating that they successfully tested 100% biodiesel in Kahe at the full load of 90 MW. Tests will be continuing for the next several months. The next test will demonstrate 24-hour operation on biofuels. One of the major observed results of the tests is that the use of biodiesel reduced opacity, NOx, and criteria pollutant emissions below the levels associated with ultra low sulfur diesel fuel[55] .
2. Maui Electric Company (MECO) uses biodiesel to start up and shut down two of its generators. MECO is also conducting tests to determine the long term impact of using biodiesel on heat rates, emissions and engine reliability.
3. HECO's Campbell Industrial Park Generating Station, Combustion Turbine Unit 1, is a new (2009) 110 MW combustion turbine unit

designed to run on 100% biodiesel. Renewable Energy Group of Iowa presently has a contract to supply three to seven million gallons of biodiesel per year. All biofuel procured by HECO must meet stringent sustainability guidelines.

HECO has determined that the use of biodiesel is its most cost-effective option for complying with Hawaii's renewable energy targets.

Biodiesel is made through the transesterification of animal fats or vegetable oils. The feedstock can come from waste sources, such as grease from restaurants, or from virgin sources, such as oil seeds, coconut, palm, jatropha, or algae. In Hawaii, biodiesel is procured from the mainland and Malaysia, though long term plans call for production of feedstocks locally.

The major issue to evaluate whether biodiesel makes sense for Guam is supply and cost. The amount of waste grease produced in Guam is not known as there is no known inventory of fats, oils, and greases. However, anecdotal evidence suggests the figure is on the order of 10,000 gallons per month[56]. There are also some concerns over shelf life, with some evidence suggesting biodiesel can begin to break down after six months. This concern can be addressed by ensuring that the supply is evenly matched to potential demand.

Also, there are significant costs associated with converting vegetable oils to biodiesel. The process requires methanol and lye, both of which would need to be imported from the mainland. There is also a question over economies of scale, and it is likely that the supply of feedstock in Guam is too limited to warrant construction of a biodiesel plant.

There is limited agricultural land available for in Guam for production of commercial quantities of oil. Dr. Marutani from the University of Guam has been conducting research on the oil yields of Jatropha and Calophyllum and has planted test plots of these species. The research is on-going and focused primarily on the quality of the oil. Research on production costs, available land and potential yields has not yet been undertaken.

While not likely to be a significant source of fuel for Guam, the use of waste oil in boilers and engines should be explored further, as should the economics of producing oil from local plants.

3.5.3. Landfill Gas

There are three potential locations on Guam where LFG-to-energy might be possible. The sites include Ordot Dump, the DOD landfill on Anderson Air Force Base, and the landfill on the U.S. Navy base at Apra Harbor.

In 2004, a study on the potential for LFG generation at Ordot was conducted[57]. The study used the EPA's LandGEM model to estimate potential methane generation and decay rates over time. This was a first level analysis utilizing a number of assumptions and estimates due to lack of specific data for Guam and Ordot. The model results indicated a rapid falloff in methane production after about 10 years. Under a middle-of-the-road scenario, the authors estimated that there was sufficient gas for nearly 3 MW of production in 2007, but that by 2017 there would only be sufficient gas for 150 kW.

There is a great deal of uncertainty over the potential for LFG capture at Ordot. The authors conclude that any LFG–to-energy plant would likely need to be sized below 500 kW to have a greater chance of running beyond five years. Given that most projects need to run for at least 10 to 15 years to pay off the investment, the willingness of an investor to finance a LFG project in Guam is questionable.

The authors suggest that if the landfill were capped and proper drainage installed there could be the potential to slow down the landfill decay rate and prolong gas generation. Given the high degree of uncertainty over the data and the model results, it is likely that the most economic option will be a capand-flare system.

3.5.4. Anaerobic Digestion

As part of the military build-up, significant improvements are planned to Guam's wastewater treatment infrastructure. The designs for the new sewage treatment plants have not yet been finalized. As the design team moves forward, the potential for installing anaerobic digesters as part of the improvements should be studied. The U.S. EPA estimates that between 491 kWh and 525 kWh of energy can be produced for each million gallons of sewage flow[58].

The Northern District Wastewater Treatment Plant (NDWTP) is presently treating about 6 million gallons per day (MGD) of flow. At one point this facility did have working anaerobic digesters although these have not been used due to disrepair for about the past 10 years. The NDWTP was designed to handle about 12 MGD of flow.

The Hagatna Waste Water Treatment Plant is designed to handle about 12 MGD and presently handles about 4.4 MGD. The U.S. Navy owns and operates a treatment plant at Apra Harbor, which is designed to handle 4.3 MGD and presently handles about 2.9 MGD. The Agat-Santa Rita plant is designed to handle .75 MGD although current flows into the plant presently exceed this design capacity. There are several other smaller plants that all treat less than 1 MGD.

To understand the potential, NREL assumed current sewage flow of 15 MGD. If 500 kWh can be produced per million gallons, the current potential on Guam is for 7,500 kWh/day. Assuming a capacity factor of 90%, this equates to about 350 kW of potential capacity at Guam's waste water treatment plants. While not a large amount, the economics should be explored to see if this option warrants further development.

3.5.5. Small-scale Biomass

There may be niche opportunities within the agricultural sector for small-scale biomass or biogas utilization. This opportunity is lower priority in terms of generation potential. The University of Guam Department of Agriculture and the U.S. Department of Agriculture's (USDA) National Resources Conservation Service (NRCS) are best situated to lead future efforts in this area.

3.5.6. Production of Dedicated Biomass Feedstocks

The production of dedicated biomass feedstocks for biofuels or for use in a solid fuel plant has not been investigated in detail. Many island communities are investigating this option. For example, there are several companies and groups in Hawaii that are developing plans for locally grown biofuels production.

However, based on the initial site visit, NREL believes that growing bioenergy feedstocks on Guam will present the following challenges:

- Small land area with steep hills, rocky cliffs, and densely growing jungle
- There is not a large agricultural industry in Guam, and there are very few commercial operations
- There are a large number of small-scale farmers with acreages from 1 to 50 acres
- NRCS and local agricultural interests would like to promote food production instead of energy, given the limited agricultural land available
- Clearing overgrown forestland to establish energy plantations presents risks from unexploded ordnances; the first cost of establishing a plantation will be very expensive.

3.5.7. Next Steps

Biomass utilization and WTE should make up some component of a wider energy strategy for Guam. Biomass offers many strategic benefits, including that biofuels can fit into the existing oil based infrastructure nearly seamlessly. Other resources such as WTE, LFG, and methane from wastewater treatment plants can be used to produce baseload power and are not subject to intermittent outages as are wind and solar.

One of the next steps would be to conduct a comprehensive biomass resource assessment and analysis of opportunities. The assessment could be performed all at one time by a single entity, or the work could be split up amongst a team of researchers. Several activities would lend themselves to being performed by students and staff from the University of Guam.

Major next steps for each of the biomass sub-technologies are described below. Detailed budget estimates would need to be prepared for each category. The cost would be dependent on level of detail required.

1. WTE is a small but important part of any strategic effort to develop renewable resources on Guam. The cost to further evaluate WTE could range from $25,000 for a pre-feasibility evaluation, to $300,000+ for a detailed waste characterization and investment grade economic and technical analysis of WTE options.

- Work with the Receiver and the SWMD to obtain any additional waste composition analysis data. The 1980 study referenced herein is old, and the 2010 waste sorts are being conducted on such a small number of samples that it is not statistically valid data. The 2010 sorts are also being performed only on the residential sector, so waste from other sectors (e.g. commercial, government, industrial, etc) is not being counted.
- Evaluate other potential sources of feedstock for a WTE plant including tires, corrugated cardboard, and green matter, which are no longer allowed to be dumped in Ordot.
- Explore the development of new regulations on WTE. This is an action that the Energy Task Force could undertake or establish a sub-committee to explore. Clearly, existing laws and legal challenges by GRRP need to be included in this analysis. The need for minimum flows and tip fees of materials going to the new landfill also need to be considered.

- One potential arrangement for a WTE facility would be to locate a plant "behind the fence" of the new landfill. In this manner, tip fees could be paid at the gate which could be used for landfill bond repayment. Rather than burying the material directly in the landfill, it could be directed to a WTE plant to produce and sell energy. The residual ash and non-combustibles could then be recycled and/or sent to the landfill. Production of energy would extract additional value from Guam's solid waste and prolong the life of the new landfill by reducing the amount of material that gets buried. A detailed cost benefit analysis of this option would need to be undertaken. The cost of this analysis is not included in the cost estimate projected above.
- Evaluate local recycling efforts, markets, and options and determine potential impacts on the waste stream and the economics of WTE.
- GPA should consider developing a separate WTE RFP once legal and regulatory issues have been resolved.

2. The production of, or use of biodiesel or vegetable oil warrants further investigation. The state of Hawaii is using both locally produced and imported biofuels to help meet its renewable energy targets. The cost impact of this analysis is estimated to be $25,000 - $50,000, excluding the costs of a direct test in GPA's engines. The cost for that would need to be ascertained based on HECO's experiences.

- Conduct a survey of waste grease supplies on Guam. Determine who produces the material, how it is collected and disposed, what the costs are, and whether there are any opportunities for a local business to manufacture biodiesel locally.
- Determine the expected costs of procuring biodiesel from the mainland or Asia.
- Gather case study data associated with the use of biofuels or vegetable oil in diesel systems in Hawaii, other islands, or other communities.
- Determine equipment modifications needed to use biodiesel or vegetable oils in the GPA system. Based on other case studies or work done in Hawaii, it would be useful for GPA to conduct a test of biofuel use within its system. This would be a comprehensive assessment that evaluates costs, impacts on heat rates and performance, fuel storage and handling, air emissions, and overall feasibility.

- Estimate the levelized cost of energy produced from biodiesel in various blends of biodiesel/petroleum diesel percentages.
- Prepare a report and briefing for GPA and the Energy Task Force to communicate results of the detailed assessment.

3. Based on the previously-mentioned study by Duenas Associates and URS, the production of energy from LFG from Ordot does not appear to be very promising. Estimated cost of these activities is less than $5,000.

- Gather additional data on the potential for Ordot to produce and capture landfill gas post closure, as appropriate. This could be accomplished through a brief meeting with the Receiver and SWMD.
- Meet with Naval Facilities Engineering Command (NAVFAC) to discuss DOD landfills, and whether they have any plans for assessing methane potential at on-base landfills.
- Monitor GPA RFPs for any possible LFG bids.

4. The additional cost of the analysis of this option would be $25,000 for a preliminary assessment to $100,000 for a detailed feasibility study.

- Determine who has responsibility for the design of wastewater treatment plants associated with the military build-up. Meet with the responsible parties to discuss present construction plans and proposed wastewater treatment technologies. Identify whether it is possible to evaluate the economic and technical feasibility of adding anaerobic digestion to the wastewater treatment plants prior to final system design and capitalization.

5. While there may be challenges to growing dedicated biomass for energy, it still makes sense to explore this option further. The following steps are recommended. The cost to perform this analysis is estimated at $25,000 to $50,000.

- Using geographical information systems, filter out all steep lands and developed areas to identify areas that are simply not suitable for biomass feedstock production.
- Overlay land use and vegetation maps to determine current usage of the land base.

- Evaluate land ownership of large contiguous parcels. The objective would be to identify potential lands suitable for growth of biomass feedstocks. DOD is likely to have the land—whether it is available for purposes of biofuel production would need to be determined.
- Estimate the acreage available as well as potential yields of various species.
- Evaluate technical and economic feasibility of producing biofuels or solid feedstocks.

4. Summary of Recommendations and Conclusion

The site visit and subsequent research on Guam's energy efficiency and renewable energy opportunities is a basic, preliminary overview to provide the Energy Task Force with a set of next steps. The areas of opportunity that have been identified are based on this preliminary review. Further investigation or consultation may be required before undertaking any next steps. The key opportunities, potential steps for policy, and concluding points are summarized in this section.

4.1. Key Opportunities

Examples of energy efficiency and renewable energy opportunities to include in strategic planning or for further consideration are summarized below.

General

- Based on existing overall baseline energy use, develop specific goals to reduce overall fossil fuel consumption.
- Develop a Strategic Energy Plan and Implementation Strategy to implement near-term opportunities identified in this report, and to develop approaches for further feasibility analysis for identification and implementation of longer-term opportunities.
- Policies, programs, regulations, and codes/standards should be reviewed for energy efficiency and renewable energy targets. High standards should be set to push the industry in the direction of reducing overall fossil fuel consumption.

- Conduct a transportation assessment to determine energy useage, analyze options, and identify reduction and renewable strategies.

Energy Efficiency

- Work with technical experts to design educational materials for Guam to address specific information and outreach needs. Flyers, posters, and stickers can be designed that could be distributed among the appropriate building sectors.
- Develop training courses and materials for both architects and workers in the construction sector to improve energy awareness. Energy efficiency in retrofit, renovation, and new construction can be included in terms of design, materials, and equipment.
- Create guidebooks and self-assessment checklists. These guidebooks could be designed as standalone products for specific sectors or integrated into the training courses described above.
- Create a cool-roof program by identifying potential buildings that can be used as pilot projects.
- Continue existing energy audits with private companies and utilize the results to benefit the entire building portfolio.

Renewable Energy

- Conduct and analyze results from wind anemometry studies to capture wind potential.
- The Energy Task Force should continue to work with GPA to identify potential sites for solar PV installation.
- Develop solar thermal strategies for residential, commercial, and governmental buildings
- DOD and GPA should continue to work cooperatively on the drilling of geothermal temperature gradient holes.
- Consideration should be given to all alternative fuel potential in the form of WTE, LFG, biofuels, and biomass crops.

4.2. Conclusion

There are many opportunities found in Guam for the adoption of energy efficiency and renewable energy technologies, and development and implementation of energy policy, standards, and codes that has been outlined in this report.

Due to fluctuating energy prices, including the record high price for oil in 2008, and the imminent military build-up, Guam has focused attention on the need to develop alternatives to its near 100% reliance on fossil fuels. Guam is in a unique position to undertake a concerted effort to address current energy and environmental challenges. Due to its geographic isolation, non-renewable resources are extremely limited and subject to large volatility in pricing and availability. Energy security is fundamental to Guam's economic future and sustainability. Consequently, creating a stable investment atmosphere remains challenging, while the quality of life for residents and visitors continues to be affected. However, the unique position of Guam also offers many natural advantages. The island is endowed with diverse sources of alternative energy.

With NREL's assistance, an Energy Task Force has been established and the following mission and vision of the have been adopted:

Vision

Create a sustainable future for Guam by reducing dependence on fossil fuels.

Mission

Unite the citizenry through the vision and enhance the well-being of Guam's population, ensure national and economic security through energy independence and diversification, improve the local environmental quality, decrease Guam's vulnerability as a small island state, and further the development of the territory by developing a comprehensive energy plan that will embrace energy efficiency and conservation.

In conclusion, a strong energy plan surveys and assesses need and provides analysis and recommendations for appropriate technologies and programs relevant to its environment while addressing economic viability. It addresses a full range of energy options, with a focus on energy efficiency and renewable energy technologies and lays out a road map for socioeconomic development.

Plans to develop an implementation plan are next steps once an energy strategic plan has been completed. A plan is only part of the solution—it lays the groundwork. Action strategies will develop the roadways in which these vehicles can drive energy efficiency and renewable energy market adoption and deployment.

A. Utility Baseline Information

A.1. Generation

GPA must meet or exceed the following generation performance standards:

- 90% or greater of generation to come from base load plants
- 10% or less of generation to come from Combustion Turbine (CT)/Diesel generation.

Table 10 lists the base load generation. Table 11 is a list of the Independent Power Producers. Table 12 is a list of the total generation resources (traditional fuels) on the island and Figure 24 shows the installed capacity over time.

Table 10. GPA Baseload Generation

Generation Unit	2003	2004	2005	2006	2007	2008	2009
Cabras Unit #1	78.1%	65.0%	75.0%	82.5%	85.0%	85.0%	87.0%
Cabras Unit #2	63.4%	94.0%	75.0%	82.5%	85.0%	85.0%	87.0%
Tanguisson Unit #1	96.4%	89.0%	85.0%	87.0%	87.0%	87.0%	87.0%
Tanguisson Unit #2	80.7%	42.0%	85.0%	87.0%	87.0%	87.0%	87.0%
Cabras Unit #3	0.0%	56.0%	62.0%	76.0%	90.0%	90.0%	90.0%
Cabras Unit #4	65.5%	67.0%	62.0%	76.0%	90.0%	90.0%	90.0%
MEC Unit #8	83.4%	95.0%	90.0%	90.0%	90.0%	90.0%	90.0%
MEC Unit #9	87.0%	96.0%	90.0%	90.0%	90.0%	90.0%	90.0%
Average Unit EAF Targets	69.3%	75.6%	78.0%	83.9%	88.0%	88.0%	88.5%
Weighted Average EAF Targets		77.4%	77.4%	83.6%	87.7%	87.7%	88.4%

Table 11. Independent Power Producers

Plant	IPP	Plant Capacity (MW)	Contract Effective Date	Contract Termination Date
Piti Unit 7 (Combustion Turbine)	TEMES	40	December 1997	December 2017
Piti Unit 8&9 (Slow Speed Diesel)	MEC	88	January 1999	January 2019
Tanguisson Unit 1&2 (Steam)	Pruvient	53	August 1997	August 2017

Table 12. GPA Generation Resources

Unit	Technology	Fuel	Capacity, MW	Service Date
Cabras 1	Steam Turbine (ST)	RFO No. 6	66	1974
Cabras 2	ST	RFO No. 6	66	1975
Cabras 3	Slow Speed Diesel (SSD)	RFO No. 6	40	1996
Cabras 4	SSD	RFO No. 6	40	1996
Piti 8 (MEC)	SSD	RFO No. 6	44	1999
Piti 9 (MEC)	SSD	RFO No. 6	44	1999
Tanguisson 1 (PRU)	ST	RFO No. 6	26.5	1976
Tanguisson 2 (PRU)	ST	RFO No. 6	26.5	1976
Dededo CT 1	Combustion Turbine (CT)	Diesel No. 2	23	1992
Dededo CT 2	CT	Diesel No. 2	23	1994
Machche CT	CT	Diesel No. 2	21	1993
Marbo CT	CT	Diesel No. 2	16	1993
Yigo CT	CT	Diesel No. 2	21	1993
Piti 7 (TEM)	CT	Diesel No. 2	40	1997
Dededo Diesel 1-4	Medium Speed Diesel (MSD)	Diesel No. 2	2.5 ea/10 total	1972
Talofofo Diesel 1 and 2	MSD	Diesel No. 2	5 ea/10 total	1994
Paluntat Diesel 1 and 2	MSD	Diesel No. 2	4.4 ea/8.8 total	1993
Tenjo Diesel 1-6	MSD	Diesel No. 2	4.4 ea/26.4 total	1994

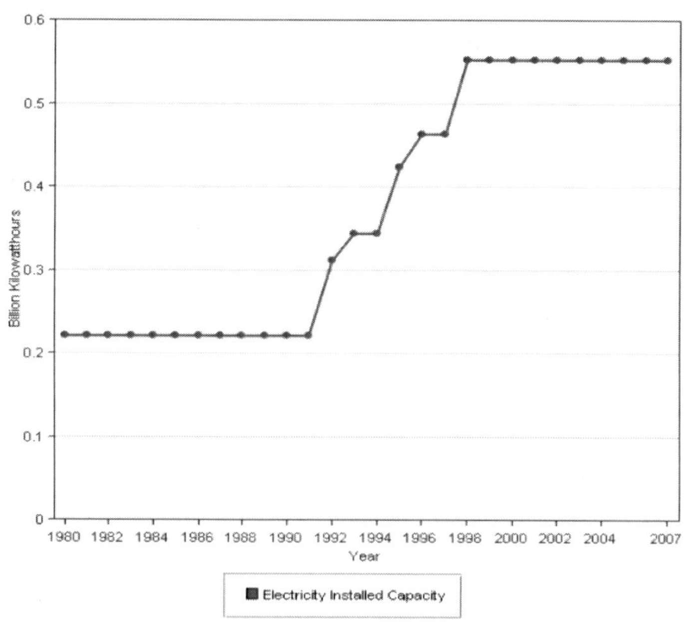

Source: EIA.

Figure 24. CPA installed capacity (in billions of kW).[59]

In 1995 and 1996, GPA commissioned Cabras #3 and #4. These units use a slow-speed diesel technology and were the largest units available in this class. This technology converted fuel into electrical energy using about a 25% to 30% less fuel than Cabras #1 & #2. The newly commissioned Cabras units #3 and #4 experienced high forced outage rates over the next two years. However, by 1997 GPA had reduced loss-of-load outage durations due to lack of generation from over 600 hours each in FYs 1995 and 1996 to about an hour.

The Cabras plant includes two 66 MW steam turbine generator units (#1 and #2), and two 40 MW slow speed (102.9 RPM) diesel engine generator units. All generator sets operate in base load service. The steam turbines are General Electric (GE) 66 MW, 3,600 RPM with GE 77,647 kVA, 0.85 pf, 13,800 volts, 3,249 amps generators, and the diesel generators are ABB, SA, type W.950/95/70, 49,280 kVA, 0.8 pf, 102.9 RPM, 13.8 kV, 2062 amps, 3 phase wye, 70 poles; air cooled. . Seawater is used as the cooling medium in the main condensers of Cabras units #1 and #2. It is also the jacket (engine) cooling medium for Cabras Units #3 and #4. All 4 generators produce electricity at 13,800 volts. The voltage is then stepped up to 115,000 volts (115 kV) in the main transformers (2 x 80 MVA for #1 and #2, and 2 x 50 MVA for #3 and #4) for transmission and distribution.

The Tanguisson plant consists of two 26 MW steam turbine generator units. Both units are operated in base load service. The turbines are GE 26,500 kW, 3,600 RPM. The generators are GE 29,412 KVA, 0.90 pf, 13,800 volts, 1,179 amps.

The Dededo combustion turbine plant consists of two GE Frame 5 machines. CT #1 (CT1) is a Model MS 5001 PA (advanced version) rated at 23 MW. CT #2 (CT2) is a Model MS 5001 P (standard version) rated at 22 MW. The units are used for peaking and emergency operations. The combustion turbines are rated 5,100 RPM (turbine), 25,000 kW. The generators are GEC Ahlstom; 26,200 kVA; 3600 RPM; 13.8 kV. Their rated outputs are 23 MW (CT1), and 22 MW (CT2). The generators produce electricity at 13,800 volts. The voltage is stepped up to 34,500 volts (34.5 kV) in the main transformers (30 MVA maximum rating) for transmission and distribution.

The Dededo diesel plant consists of four General Motors (GM) Electro-Motive Diesel (EMD) diesel engine generators. Each diesel generator is rated at 2.5 MW. The plant's total generating capacity is 10 MW. The units are used for peaking and emergency service. The diesel engines operate at 900 RPM. The generators are GM-EMD; Model A20-C1; 3,250 kVA, 0.8 pf, rated output

2.5 MW, 4,160 volts, air cooled. The engines can operate with just one fan in operation at a slightly reduced load (2.2 MW). The generator voltage is stepped up to 24 kV for transmission and distribution.

The Macheche and Yigo combustion-turbine plants each consist of one GE LM2500 CT generator unit. Each unit is rated at 22 MW. These units are used for peaking and emergency operations. The combustion turbines are 3,600 RPM, 25,000 kW. The generators are Brush Electric Model BDX7-167E, 3m600 RPM, 13,800 volts, 25,000 kW, type HC/OP/OPLTR, class OA/FA/FA, 18/24/30 MVA, 13.8 .90 pf, air cooled (rated output 22 MW). The generator voltage is stepped up to 34.5 kV for transmission and distribution.

The Manenggon plant consists of two Wartsila-ABB diesel engine generators. Each diesel generator is rated at 5 MW. The plant's total generating capacity is 10 MW. The units are used for peaking and emergency service. The diesel engines are 5,522 kW, 720 RPM. The generators are ABB Stromberg; type HSG 900 LS10; 7,250 kVA; 13.8 kV; 303 amps; air cooled. The generator voltage is stepped up to 34.5 kV for transmission and distribution.

The Marbo plant consists of one Fiat TG 16 CT generator unit. This unit is rated at 16 MW and is used for peaking and emergency operations. The CT operates at 4,914 RPM. The generator operates at 1,800 RPM, 19,000 kVA, 13.8 kV, 794.9 amps, 0.8 pf, air cooled. The generator voltage is stepped up to 34.5 kV for transmission and distribution.

The Tenjo plant consists of six Caterpillar-Kato diesel engine generators (two units are currently being overhauled). Each unit is rated at 4.88 MW each. The units are used for peaking and emergency service. The diesel engines are Caterpillar; Model 3616, 6,095 HP, 16 cylinder, V-type, turbocharged, 900 RPM, #2 fuel (diesel) oil. The generators are Kato: Model A256730000, 4,880 kW, 6,100 kVA, 0.8 pf, 13.8 kV, 255 amps, air cooled. The generator voltage is stepped up to 34.5 kV for transmission and distribution.

A major project aimed at reducing forced power outages was completed after two years. The project focused on automating generation controls, which allows for other generators to increase power output in the event one of the larger base load units trips and is forced offline. These forced outages caused power outages in many areas while reserve generators were being brought online to restore power. The upgrade enables standby generators to quickly make up for the loss of generation through automation instead of coordinating a restoration of power through the slower manual processes.

GPA increased the number of Fast Track generating units online. Work at three 20 MW combustion turbine units (Dededo CT No. 2, Macheche, and Yigo CT) resulted in an additional 60 MW of reserve that can be brought online to support maintenance, overhauls, or emergency shutdowns at other GPA base load generating units.

The Fast Track units represent one part of the generating mix that is used to provide recovery and support during typhoon restorations and blackouts. Bringing these units online adds operational flexibility during hours of peak demand today and anticipated additional load from the pending military buildup tomorrow.

The Fast Track units are a component in the project that connects large load centers to major substations via underground transmission lines. The Dededo CT Plant is connected to Andersen Air Force Base via an underground transmission line, the Yigo CT is within close range of Andersen Air Force Base, and the Macheche CT is connected to the Guam International Airport Authority.

Figure 23 in Section 3.3.2 of the report shows a map of the island's transmission system as well as potential locations for solar and wind systems. This indicates where GPA could collaborate with PV and wind installers to tie into the existing electric grid.

A.2. GPA System Improvements in FY 2009

A.2.1. Major Transmission Lines Secured Underground

Electricity is delivered through a transmission and distribution grid consisting of a combination of 115 kV, 34.5 kV, and 13.8 kV lines. GPA energized the final portion of the massive project to move main transmission lines underground (as shown in Figure 25), where they are shielded from typhoons and other disasters and contribute to timely recovery of power and other services for the entire island). Funded by a grant from the Federal Emergency Management Agency, the project cost $25 million and took four years to complete.

The final leg of the project was completion of the Harmon to Macheche underground transmission line, which provides a secure, direct line from the Macheche Power Plant to the Guam International Airport Authority. Since 2005, GPA placed major transmission lines underground that serve the Guam Memorial Hospital and central areas of Harmon, Tamuning, and Tumon.

GPA also buried a significant portion of line connecting the Tanguission Power Plant to the Harmon Substation. The Tanguission Power Plant is capable of producing over 50 MW—this number added to the 20 MW from Macheche Power Plant means that GPA can potentially transmit 70 MW of uninterrupted power during tropical storms or typhoons via the underground network.

Avoiding repairs on the underground lines is a savings to rate payers that protects revenue and restores island power to residences and businesses after a major storm. The hospital and airport will have the potential for uninterrupted power during tropical storms and typhoons. In addition, GPA's restoration crews will be able to work on restoring service to neighborhood homes, schools, and businesses if lines to major substations, the hospital, airport, and businesses and hotels along Tumon Bay are sheltered from storm damage.

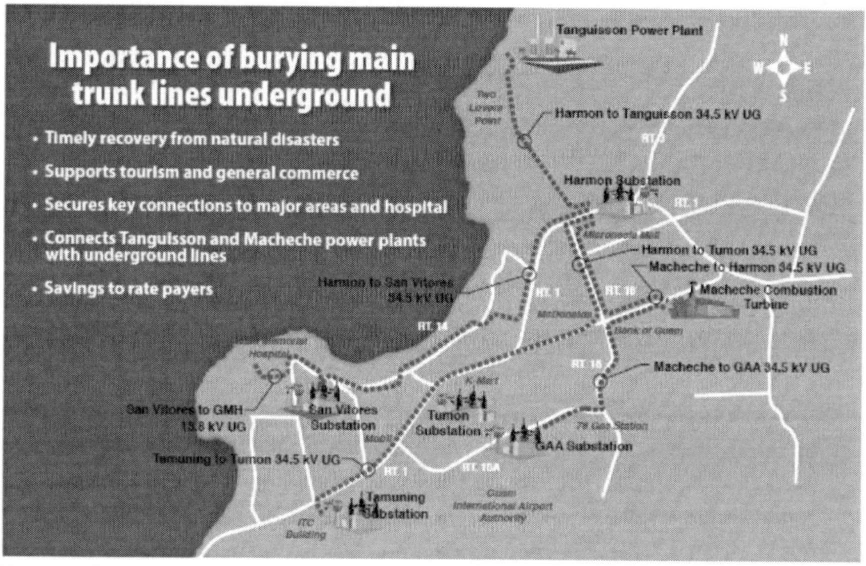

Source: GPA.

Figure 25. CPA buried transmission lines.

A.2.2. Net Metering

The first use of "net metering" was implemented at a residence in Barrigada in 2009. Net metering allows the GPA customer to offset power consumption by generating renewable energy, such as solar power. The renewable power incentive program was made possible by passage of PL 27-132 and adoption by the PUC. The legislation was intended to encourage the use of new power generation technologies to defer the need for additional capital investment by GPA.

A.2.3. Groundwork Laid for Pre-Paid Meters

The cost of residential electricity is about $0.23/kWh, where $0.15 is the cost of fuel and $0.08 is for infrastructure. A pilot project to determine the feasibility of pre-paid meters for residential customers was launched in 2009. One hundred households—primarily those of GPA employees—volunteered to participate in the project, which is spearheaded by the GPA Engineering Department. The technology used in the project has features that easily interface with GPA's existing computer systems.

In the future, pre-paid meters will allow residential customers to manage their electric utility accounts. Families will be able to manage their accounts by monitoring consumption and conservation of electricity. Instead of monthly bills and security deposits, customers will be able to choose the payment schedule that is best suited for their personal needs. If the pilot project is successful, GPA will offer prepaid meters to more residential customers.

A.2.4. Declining Sales

GPA's sales have been trending downward for the last few years as is shown in Figure 26. While the number of customers is growing, the average usage is declining as a result of conservation efforts by many of GPA's customers. These conservation measures are the result of the steep increases in the fuel portion of the bill for the last few years. GPA expects to see positive growth for FY 2011 as conservation measures become fully realized, the impacts of the military buildup begin to show, and the economy begins to recover from the impact of the world-wide economic crisis.

A.3. Utility Energy Efficiency

GPA staff came up with PMCs to improve base load plant reliability and efficiency. GPA staff recognized that GPA did not have sufficient plant management, technical, and plant operation acumen at GPA to run its base load facilities well. Keeping many of these skill sets full-time at GPA is economically prohibitive.

Additionally, GPA already had difficulty recruiting to fill technical and professional positions. Also, the Government of Guam's procurement practices do not support an operations environment well as procurement issues often result in prolonged unit outages. Furthermore, GPA recognized the need for better, consistent training of its plant staff. Finally, staff foresaw that performance-based compensation would best drive exemplary performance and better protect the ratepayer from poor performance.

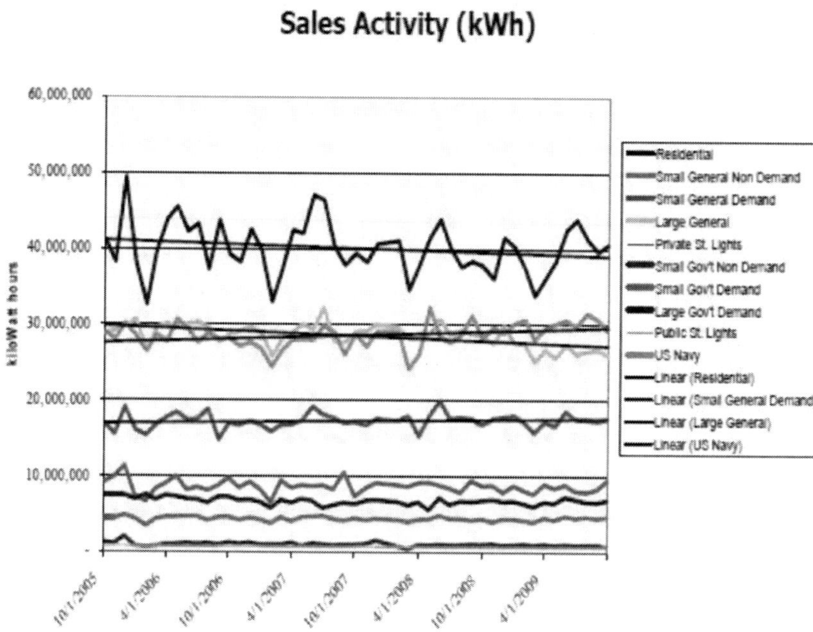

Source: GPA.

Figure 26. Decline of CPA sales.

Peaking unit technologies are relatively inexpensive and quick to install but expensive to operate. Therefore they are ideally operated only during system peak demand periods or as reserve units in the absence of reliable base load capacity. Efficient base load units require much longer permitting and construction lead times. However, they possess much higher capital requirements for installation but are less expensive to operate. Intermediate units have unit characteristics between peaking and base load.

Guam's year-round tropical climate and tourism-based economy results in a relatively flat load cycle with high load factor. Such characteristics tend to favor base load technology additions since operation near the peak is the norm (flat load profile). As an example, Figure 27 shows the GPA average demand hourly load shape for the period April 29–30, 2008. Note that GPA requires peaking capacity for only four hours for about 15 MW to 19 MW incremental peak.

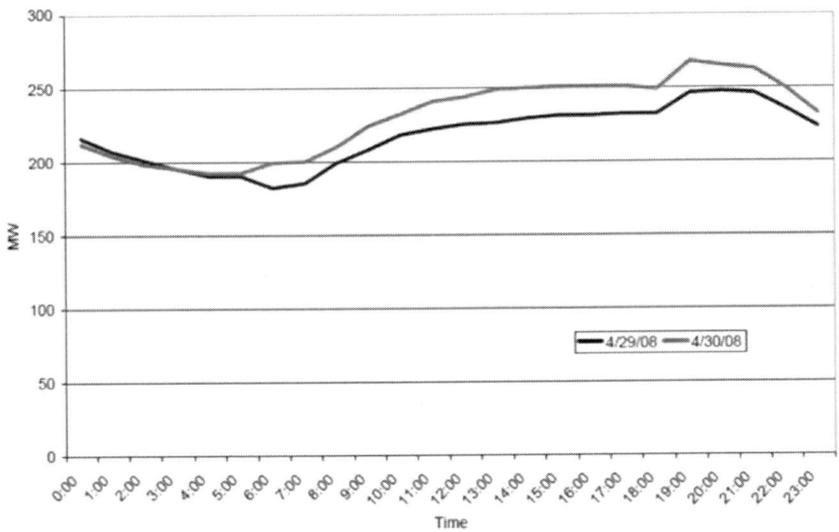

Source: GPA.

Figure 27. CPA average hourly generation requirements.

Regarding the distribution system, savings would be realized by improving technical efficiency (improved connections, reducing line resistance, correctly sizing transformers to their loads, etc.) as well as improving nontechnical efficiency (inaccurate billing, meters not being calibrated, etc.).

B. Energy Efficiency Baseline

Sectors of energy consumption are categorized by the Secretariat of the Pacific Community and GPA as commercial and government, residential, U.S. Navy, and street and outdoor lighting.

Within the commercial and government sectors we can include the hospitality industry, medical, government, commercial, and higher education. The construction of buildings on Guam is consistent with tropical island building code, which emphasizes construction methods to withstand typhoons typical of the area. From observations, the majority of commercial and residential buildings are reinforced concrete buildings with high air conditioning loads. The heat and humidity levels dictate the use of air conditioning on island. Some buildings are constructed of concrete or wood with tin roofs, but seem to be mostly residential. These buildings are not preferred due to their lack of withstanding typhoon conditions. Glazing is either single or double-glazed with external typhoon shutters attached to the building to be closed during inclement weather.

B.1. Current Status

The Guam Energy Office has identified measures to improve energy efficiency on the island; Recovery Act funding was requested to implement many of these tasks. Certain activities were entered into memorandums of understanding or RFPs, including the following:

- Improve municipal lighting
  - Replace old metal halide and high pressure sodium light bulbs along highway and major suburban roadways with energy-saving Light Emitting Diode (LED) lights
- Retrofit public facilities
  - Energy audits performed by energy services companies at public facilities, such as schools, housing, and community centers to identify potential retrofits. Facilities will be retrofitted based on audits.
- Financial incentives for building improvements
  - Rebates for residents, non-profit organizations, and businesses for energy efficiency measures, including solar hot water heating, lighting, insulation, air conditioning, and windows.
- Resource conservation manager program
  - Reduce government expenditures for energy, water, fuel oil, solid waste disposal, and other utilities by improving energy use practices, auditing utility billing and services, and installation of resource management equipment.
- Residential energy performance score
  - Home energy audits to identify opportunities for efficiency improvements and encourage voluntary retrofits by providing information to the real estate market. Members of Guam

Contractors Association, building code inspectors, and GPA staff will be trained to deliver the audits.

- Community education and outreach
    - Guam Department of Education will work with island schools and other organizations to increase energy awareness and promote future conservation of resources. Also, education curriculum and courses in Energy Technician Training and Building Code compliance will be developed.
- Sustainable Development Forward
    - Energy saving guidance for households, database to evaluate savings estimates for residential and commercial facilities, and adoption of state-of-the-art building energy codes that will increase energy efficiency in new buildings by 30% or more.

Recovery Act funding was requested and granted for the state energy program, weatherization assistance program, energy efficiency conservation block grants, and the energy efficiency appliance rebate program. This funding assists with implementing a few of the measures outlined by the Energy Office, however does not provide funding for all measures identified. A summary of the program and the funding allocation for each area is summarized in Figure 28[60].

Program	State Energy Program	Weatherization Assistance Program	Energy Efficiency Conservation Block Grants	Energy Efficiency Appliance Rebate Program
Award (in millions)	$19.1	$1.1	$9.6	$0.2
	The Guam Energy Office has been granted $19.1 million in State Energy Program funds to invest in territory-level energy efficiency and renewable energy priorities.	The Guam Energy Office has been granted $1.1 million in Weatherization Assistance Program funds to scale-up existing weatherization efforts on the island, creating jobs, reducing carbon emissions, and saving money for Guam's low-income families. Over the course of the Recovery Act, Guam expects to weatherize more than 200 homes. The program also includes workforce training and education as part of the island's efforts to develop a green workforce.	Guam has been granted $9.6 million for Energy Efficiency and Conservation Block Grants (EECBG) to develop, promote, implement, and manage local energy efficiency programs.	The Guam Department of Administration has been granted $166,000 for the Energy Efficient Appliance Rebate Program, which offers consumer rebates for purchasing certain ENERGY STAR® appliances. These energy efficient appliances reduce energy use and save money for families, while helping the environment and supporting the local economy.

Source: DOE.

Figure 28. Recovery Act funding allocated to energy efficiency measures on Guam (in millions of U.S. dollars).

While the Energy Office has determined areas of focus and has been granted funding to implement specific measures the work has only just begun and there are gaps in the level of coverage on-island. NREL's goal in assessing the current level of energy efficiency is not intended to provide duplication of efforts, but rather to assist the island in identifying areas that have not been noted previously for energy conservation measures.

B.2. Initial Observations by Sector

B.2.1. Commercial and Government

Of all sectors on-island the commercial and government buildings seemed to have the most comprehensive implementation plan. During the island site visit, investment grade energy audits were being conducted at the airport and five hotels, new construction at Guam Community College was being built to achieve LEED certification, and community offices at Dededo were undertaking low-cost measures toward energy reduction. NREL staff were not able to visit any medical buildings to observe current practices, however this may be an area to explore further with investment-grade audits.

In general, observations and interviews within the commercial and government sector indicated an overall need for energy awareness. Simple, low-cost measures could be implemented to increase energy efficiency and conservation on island. Examples of this include temperature settings on air conditioning units/systems and better lighting controls.

B.2.1.1. Temperature

The Dededo Senior Center is a good example of temperature control within a community building, and simple improvements could be made. The thermostat was set to 65°F on a forced air system with ceiling vents. The temperature at ceiling level was 2°F cooler than at floor level, indicating striation. The air conditioning was turned on, windows were open in the bathrooms, and gaps between the glass doors allowed cool air to leak out of the building as shown in Table 13.

Low-cost measures that could be implemented at the Dededo Senior Center include the following:

- Add vent guards to the ceiling vents to direct air downward to cool the occupants below

- Close windows and add door foam or brush seals to prevent cool air from leaking outdoors (gaps are shown in Figure 29).
- Set the thermostat to a higher temperature, gradually increasing it over a few months to 74°F.

ASHRAE states that there is only a 10% dissatisfaction rate for temperatures between 74o-78o F61 . The challenge is creating a culture change where people are encouraged to dress appropriately for external conditions and not expect buildings to be cooled to ~60° F.

Table 13. Dededo Senior Center Indoor Temperature and Related Costs

Dededo Senior Center	
Hours of Operation	8am – 8pm
Cost/kWh	28¢/kWh
Monthly Consumption	1400-1700 kWh/month
Thermostat setting	65°F
Temperature gradient	2°F between floor and ceiling

Source: NREL/PIX 18584.

Figure 29. Visible gaps in door at Dededo Senior Center.

Humidity control in hot, humid climates prevents mold growth, damp rot in building structures, and health and comfort problems among occupants. Unfortunately, conventional air conditioners cannot control temperature and humidity independently of one another. ASHRAE recommends a 60% relative humidity maximum at a temperature setting of 78° F. Simple actions can be taken to lessen the need for dehumidification within a building, including:

- Do not set thermostats to 'fan on' position as this will blow moist air whether the cooling system is on or not
- Use exhaust fans in bathrooms, utility rooms with dryers, and kitchens to direct moist air outdoors.

Higher-cost options for community buildings like the Dededo Senior Center include measures such as improvements to the air conditioning system or creating a cool roof. Modifications at this level should be done during major renovation as the capital costs will be lower than if it is a stand-alone project.

- When replacing an air conditioning system consider a modern desiccant cooling system, which removes humidity at a lower rate of energy consumption
- Retrofit an existing roof into a cool roof by using reflective materials to lower the heat absorption on the roof.

Desiccant cooling is a technology that is advancing and systems are available on the market. These systems typically use water as a coolant and silica gels or lithium-chloride inside desiccant wheels to dehumidify. There are two types of desiccant systems, either solid or liquid. The solid systems can be utilized with a solar heating system that acts as a heat exchanger, however, these systems often need special design modifications in hot, humid climates. Standard configurations of the desiccant cycle are not typically able to reduce humidity low enough for direct evaporative cooling. Liquid systems have shown higher air dehumidification rates at the same operational temperature as solid desiccant systems.

A cool roof is defined as a roof that has a lower roof temperature than traditional roofs, while the sun is shining. Cool roofs are designed to reflect sunlight and emit heat more efficiently than typical roofs through solar reflectance and thermal emittance. Dark roofs absorb approximately 90% of solar energy and the temperature on the roof can reach up to 150°F. Higher roof temperatures increase the internal temperatures, which impacts the demand on the air conditioning system. Light colored roofs, on the other hand, absorb less than 50% of the solar energy, reducing roof temperature and air conditioning loads.

Cool roof materials vary from coatings that can be rolled, brushed, or sprayed to tiles and shingles. Before any cool roof is applied, the roofing structure and integrity should be checked and leaks or problems should be repaired. Coatings are available in acrylic, silicone, and urethane, as summarized in Table 14. Coating varies by manufacture and improper application can cause flaking. Installers should follow manufacturer-specific recommendations.

Table 14. Summary of Cool Roof Coatings by Type

Coating Type	Properties	Cost
Acrylic	Water based, easy to handle, good adhesion to most roof types, most commonly used reflective coating, cures by evaporation, reasonably strong, very sensitive to weather	Moderate
Silicone	Solvent based, typically used for spray polyurethane roofs, weather very well, weaker tensile strength, good water resistance	More expensive
Urethane	Solvent based, 3-10 times stronger than acrylic, greater adhesion to most roof types than acrylic, most cure with exposure to air, less sensitive to weather, more difficultto work with	More expensive

Tile roofs can be designed as cool roofs by selecting tiles with specific reflectance and emittance qualities. Permanent or unfading tiles are often preferred as they retain their surface properties and can yield better lifetime energy savings. With all cool roofs, maintenance is essential to prevent build-up from pollution, debris deposited by wind and storms, mold or algae growth. Any of these naturally occurring deposits will decrease reflectance, leading to a less effective cool roof.

More information about cool roofs can be found on DOE's website on cool-roof resources[62].

B.2.1.2. Lighting

Lighting efficiency could be improved through the use of more efficient technologies and energy conservation techniques. Observations indicated the use of less efficient bulbs/lamps and a lack of sensors or awareness in turning lighting off when daylighting was adequate and did not require the use of additional lighting.

Figures 30-32 are examples of adequate daylighting levels without the need for internal lighting. Figure 30 and Figure 3 1 are examples of effective daylighting levels without the need for internal lighting, again, on an overcast day. Figure 3 0 is a photograph of the museum at the Lette of Freedom with the lighting on and Figure 3 1 is a photograph of the same museum with the lighting switched off. There is no noticeable difference in the light level readings in foot candles (fc) with the lights on or off. When NREL staff entered the museum it was unoccupied and all lighting had been left on.

The hotel lobby in Figure 32 has large windows which, even on an overcast day, allow enough light in the lounge areas of the lobby. The red circles emphasize the lights that are turned on, but are not effective in providing additional lighting. Hotel staff were asked if the lights could be switched off, however their investigation into the situation indicated the lighting system is wired in such a way to prevent isolation of specific banks of lighting. The entire lobby seems to be on the same circuit, which does not facilitate control of lighting during daylight hours.

Source: NREL/PIX 18586.

Figure 30. Museum lights off at the Lette of Freedom.

Source: NREL/PIX 1858.

Figure 31. Museum lights on at the Lette of Freedom.

Source: NREL/PIX 18585.

Figure 32. Lobby lighting at the Tumon Bay Hyatt.

Hotel lighting would benefit from daylight sensors and wiring that allow lighting to be isolated and switched off when daylighting is adequate. Public areas, such as the museum, would benefit from motion sensors to turn lights on when people approach the museum and switch off if the space is unoccupied.

Effective lighting operation through the use of controls (timers, motion sensors, or daylight sensors) can conserve energy. Lamp type also impacts lighting energy consumption through the light level output and the amount of energy required to generate the light.

Lighting technology has improved significantly in the last decade, evident in the options of higher efficiency lamps. One example is the shift in use of tubular fluorescent lamps from higher wattage lamps to lower wattage lamps. T12 lamps are tubular fluorescent lamps with a larger diameter than either T8 or T5 lamps. The T8 and T5 lamps are lower wattage bulbs. Generally T8 lamps can directly replace T12 lamps, whereas T5 lamps may need an adaptor for the fittings on either end, or an entirely new ballast. Savings of up to 10% are achievable by replacing a T12 lamp with a T8 lamp in switch-start luminaires.

Other technological advancements have been made in direct replacement bulbs for other lamp types. Compact fluorescent lamps act as direct replacements with lower wattage for incandescent bulbs. LEDs have the highest efficiency of all lamp types, but are not appropriate for all applications. LEDs generate very little heat, but are most effective in directional applications, such as halogen bulb replacements for retail displays. These technologies are becoming less expensive and new applications are being designed at a quick pace. This is a market to monitor for energy efficiency and lighting design within retrofits and new construction.

Table 13 summarizes the recommended lamp types by application, either internal or external and the space that is being used.

B.2.2. Residential

The NREL team was not able to assess the residential sector in great detail; however this sector may be one where low-cost measures can be implemented easily through an awareness campaign in each village.

Figure 33 is one example of the type of housing found on island. Interviews on-island indicated a general lack of energy awareness despite high monthly electricity bills.

In Guam the highest energy consumption in the residential sector is water heating, followed by air conditioning, and household appliances.

Table 13. Recommended Efficient Lamp Types by Space Use

	Space Use	Recommended Lamp Type
Internal	Offices	Triphosphor tubular fluorescent, compact fluorescent, halogen replacement lamps (LED)
	Hotels	Triphosphor tubular fluorescent, compact fluorescent, LED
	Hospitals	Triphosphor tubular fluorescent, compact fluorescent
	Retail	Metal halide, white sodium, compact fluorescent, LED, halogen replacement lamps (LED)
	Emergency exit signs	LED
External	Parking lots	High pressure sodium, metal halide, compact fluorescent
	Floodlighting	Metal halide, high pressure sodium
	Feature lighting	Metal halide, high pressure sodium, triphosphor tubular fluorescent, LED
	Sports facilities	Metal halide

Source: NREL/PIX 18583.

Figure 33. Residential housing in Dededo.

Areas of improvement suggested by island residents during the October 2010 site visit included increasing knowledge of energy conservation within the construction industry. Working with local architects to improve standard building practices, reviewing the building code for energy efficiency specifications and ensuring only energy-efficient equipment and appliances are imported to Guam would contribute to the overall conservation on island.

- Construction and design practices that would be beneficial include using appropriate insulating and reflective building materials (while avoiding reflecting sunlight and heat to neighboring buildings), utilizing external window shading to prevent solar gain through glazing, installing air conditioning systems with an efficient SEER, and installing lighting systems with higher efficiencies.
- Controlling the electrical equipment imported to Guam is another area that would be beneficial. If the industry specified an efficiency standard of equipment and appliances on-island, greater efficiency would result.

While construction improvements may be possible through the USDA's Rural Development Housing and Community Facilities Program[63] or increasing energy efficiency specifications within the tropical island building code, there are basic measures that can be implemented by residents and communities.

An awareness campaign could be carried out through the local schools and village mayor's offices to encourage residents in each village to reduce energy consumption. Basic measures, such as switching off lights and air conditioners when they're not being used, raising thermostat temperatures on air conditioners, and switching to more efficient light bulbs and appliances, are all actions that can be taken within communities. Teaching children to be more energy conscious in schools is often an effective approach to reducing energy consumption within homes. Friendly competitions could be held between each village and school to encourage implementation. One option is to pilot such a program in Dededo, where the mayor and staff are enthusiastic about energy conservation and where the population is largest. The potential for energy savings is highest and could also set a precedent among other villages and mayor's offices.

B.2.3. Outdoor and Street Lighting

Outdoor and street lighting does not consume the largest portion of Guam's energy, but there are obvious measures that could be undertaken to reduce this sector of energy consumption.

The NREL team observed that outdoor lighting on buildings and street lights were left on during daylight hours when lighting was not needed. On

NAVFAC property a call was suggested to turn off a building's external lighting and island residents confirmed that street lighting control was problematic.

Source: NREL/PIX 18589.

Figure 34. Example of street lighting on during daylight hours in Tumon Bay.

The two measures that would be beneficial to this sector are improving the lamps themselves and improving controls.

- Energy-efficient outdoor lighting is available that will reduce the wattage while providing the same quality of light, but should be chosen for specific applications, as summarized in Table 16.
- Lighting controls should be installed and maintained to either switch lights off during daylight hours, be set for specific operational times through the use of timers or use motion detectors to provide light when movement is detected by humans (or in some cases, animals, depending on the application).

Table 16. Recommended Efficient Lamp Types for Outdoor and Street Lighting

	Space Use	Recommended Lamp Types
External	Street lighting	Induction lighting could be trialed, LED, high pressure sodium
	Parking lots	High pressure sodium, metal halide, compact fluorescent
	Floodlighting	Metal halide, high pressure sodium
	Feature lighting	Metal halide, high pressure sodium, triphosphor tubular fluorescent, LED
	Sports facilities	Metal halide

B.2.4. U.S. Navy

The U.S. Navy consumes approximately 20% of the island's total energy. While this is an area where improvements could be made, NREL has not included recommendations in this study due to the fact that the U.S. Navy has undertaken numerous studies and is currently in the process of implementing energy conservation measures and renewable energy. All military bases on Guam could benefit the island as a whole by ensuring that energy efficiency is a priority during the proposed military build-up, not only for its own facilities and personnel, but also for the facilities that will accommodate the construction workers who will live on-island during the buildup process. Temporary housing is often constructed with the intent of dismantling the housing once it is no longer needed, however in many cases temporary housing continues to be used for many decades and becomes more permanent than initially anticipated. As such, the military could influence the island's energy loads by working with contractors to encourage, and where within operational control, specify those additional facilities apply energy efficiency standards. One area for improvement is temporary housing and water treatment plants. Water treatment is discussed in other areas of this report.

C. Renewable Energy

C.1. Solar PV

C.1.1. Overview

This section contains discussion of the site parameters used in the Siting Study. These site parameters were investigated using the best available data and following the methodology described in the next section.

The island of Guam contains approximately 46,800 parcels on 133,900 acres. The northern part of the island is relatively flat and is more developed as compared to the southern part of the island. The island's electricity is generated by a number of relatively small fossil-fuel-burning power plants. GPA electricity is delivered through a transmission and distribution grid consisting of a combination of 115 kV, 34.5 kV, and 13.8 kV lines.

A suitable site for a solar electric generating facility on Guam would have the following general characteristics in order to enhance the cost effectiveness of the project:

- Adequate unoccupied area (60-80 acres)
- Flat or gradual southerly sloped terrain with minimal vegetative cover
- Close proximity to electrical transmission/distribution lines, access roads, and electrical load centers
- Favorable land-use designation or a clear indication that the land use designation can be changed to accommodate the solar project.

The following sections provide further information on the site parameters used in this analysis.

Land zoned for military use as well as land zoned as conservation was ruled out of the analysis. All other zoning designations were considered to be either suitable or relatively easy to change.

Parcels that are substantially occupied were ruled out of the analysis. It should be noted that the parcel map used is two years old and some parcels have been subdivided since this data was created. Therefore, several large parcels in the analysis below contain a number of residences, but have large areas unoccupied. In cases such as these, the unoccupied areas of the parcel were considered suitable and it is assumed that the subdivided portions of the parcel do not substantially interfere with the potential future use of the unused areas.

R.W. Beck provided footprint information of 60-80 acres per 10 MW plant, which GPA accepted as a target parameter for the basis of this memorandum. Based on this parameter, highly irregularly shaped parcels may not be suitable if their shape prohibits the existence of the 60-acre area. However, for the purposes of the IFB, the minimum size is 5 MW.

Relatively flat areas are ideal, but a slope of up to 10% was considered acceptable for this study. Anything beyond 10% was considered unacceptable.

Areas with a southern-facing slope aspect are ideal, but any slope between 90° and 270° (based on compass degrees) was considered acceptable. Any aspect north of this line was considered unacceptable.

An ideal site has close proximity to access roads. Most if not all of the sites are adjacent to roads that have access to village streets or highways. Thus, in general this site parameter was satisfied by most properties that were evaluated.

For economic reasons, an ideal site shall be located in close proximity to electrical transmission or distribution lines and also the electrical load center. A 34.5-kV feeder connection to the transmission system could be up to 15 or 20 mi. in length without incurring significant voltage drop. A 13.8-kV feeder connection to the distribution system could be up to 2 mi. in length without incurring significant voltage drop. The actual voltage drop and energy losses in the feeder line will be dependent on the wire size, circuit length, and line configuration, as well as on the magnitude of the current (amperes) flowing in the line.

Monitory considerations, such as cost of step-up transformers, switching and protective devices, and pole line construction, will also have a significant impact in determining the point of connection and system voltage.

Although GPA's preference is to have the solar power plant connect to the transmission system (34.5 kV), there are potential cost and equipment factors that may make a connection to the distribution system more advantageous for certain locations. Potential solar power plant developers will be required to undertake a System Impact Study to analyze the benefits, impacts, and necessary upgrades to the GPA transmission (34.5 kV) or distribution (13.8 kV) systems due to their proposed system(s). Figure 23 shows the existing GPA transmission and distribution systems near the proposed sites that were selected in this study. Additional information on the transmission and distribution systems can be obtained from GPA for other areas of the island, upon request.

C.1.2. Methodology

The following sections outline the step-by-step methodology used to screen all known parcels on the island. All steps below were undertaken in the GIS software program ArcInfo by ESRI, using the Spatial Analyst toolset.

C.1.2.1. Source Data and Known Data Gaps

The data used for the analysis in this report was primarily obtained from the following GIS layers:

1. Parcels (Cadastre2007.shp) - last updated March 2007
2. Zoning (ZoneMap_update.shp) - last updated March 2006
3. Digital Elevation Model (ned_35956460_GuamGCS93.shp) - created March 2009.

It should be noted that the parcel layer was obtained through existing Government of Guam GIS databases being used by GWA and other government agencies. This layer is missing lot numbers and ownership information on approximately 11% of the island due to "holes" in the data. However, these land areas were evaluated for slope and slope aspect and it was found that only two areas contain potential sites (the other "holes" consist of areas that are too steep). These lots are identified by the lot number "L0" and will require further research by any interested developers to determine land ownership and confirm the extents of the parcels because they are not in the currently available Cadastre files.

A total of 46,819 parcels are in the parcel layer, 586 of which were zoned "Military" and one of which was zoned "Conservation." These 587 parcels were removed from consideration, leaving 46,232 parcels, under the following zoning designations:

- Agricultural
- Commercial
- Industrial
- Residential.

Once inappropriately zoned parcels were eliminated, the remaining parcels were screened to eliminate all parcels less than 60 acres, leaving 134 appropriately zoned parcels. Of these parcels, none were eliminated due to highly irregular shapes/dimensions that made the parcel unusable. The remaining 134 appropriately zoned parcels that were >60 acres were of sufficient quantity to eliminate the need to search for parcels smaller than 60 acres. However, it should be noted that additional candidate sites may exist in the form of multiple contiguous parcels smaller than 60 acres. For instance, two contiguous parcels that are each 50 acres (and each matching the necessary slope and aspect criteria) could be joined to create a 100-acre parcel. However, this approach was not utilized for this initial screening.

Next, the 134 remaining parcels were placed into Tiers using the screening criteria discussed below (occupancy, percent of slope, and aspect of slope). Tier 1 represents parcels that appear to be the best suited for the intended purposes. Tier 2 represents parcels that have some problems, but are generally well suited according to most screening criteria. Tier 3 represents parcels that appear to have fatal flaws (such as being primarily occupied).

Of the 134 parcels remaining for consideration, many were occupied. Occupied parcels were divided into two categories: primarily occupied and minimally occupied. Minimally occupied parcels may have included large tracts of land with residential or other uses on their periphery. Primarily occupied parcels were placed into Tier 3. It should be noted that some occupied parcels could have lot line adjustments to move unoccupied portions into combination with neighboring parcels. However, this level of analysis was not considered in this study.

Percent of slope was calculated based on a Digital Elevation Model (DEM) with a 100-foot resolution. This means that elevation data for the entire island was divided into discrete cells, each consisting of 10,000 square feet (100' x 100'). This DEM was used to generate each cell's slope through the following process:

1. The average elevation of each cell is compared to the eight cells surrounding it
2. The degree of change in elevation from the cell to its immediate neighbors is determined
3. An average slope value (in %) is assigned to the cell based on this degree of change in elevation.

In this way, an average slope was calculated for each 10,000 sf block of land on the island. All cells with a slope value great than 10% were symbolized, while all cells with a slope value of 10% or less were not symbolized. The result is a Percent of Slope layer in which areas that are too steep (according to the site parameter criteria established above) are shaded red.

All parcels were evaluated to identify contiguous areas internal to the parcel that were at least 60 acres and had a slope value of 10% or less. Parcels that did not meet these criteria were placed into Tier 3.

The same DEM referenced above was used to generate each cell's aspect through a similar process as was used to calculate percent of slope. All cells with a northerly aspect were symbolized, while all cells with a southerly aspect were not symbolized. The result is an Aspect of Slope layer in which areas in which the slope is at the wrong aspect (according to the site parameter criteria established above) are shaded yellow.

When the Aspect of Slope layer is overlaid on the Percent of Slope layer, the result is as follows: areas that are too steep are shaded red, areas that are at the wrong aspect are shaded yellow, while areas that are both too steep and at the wrong aspect are shaded orange.

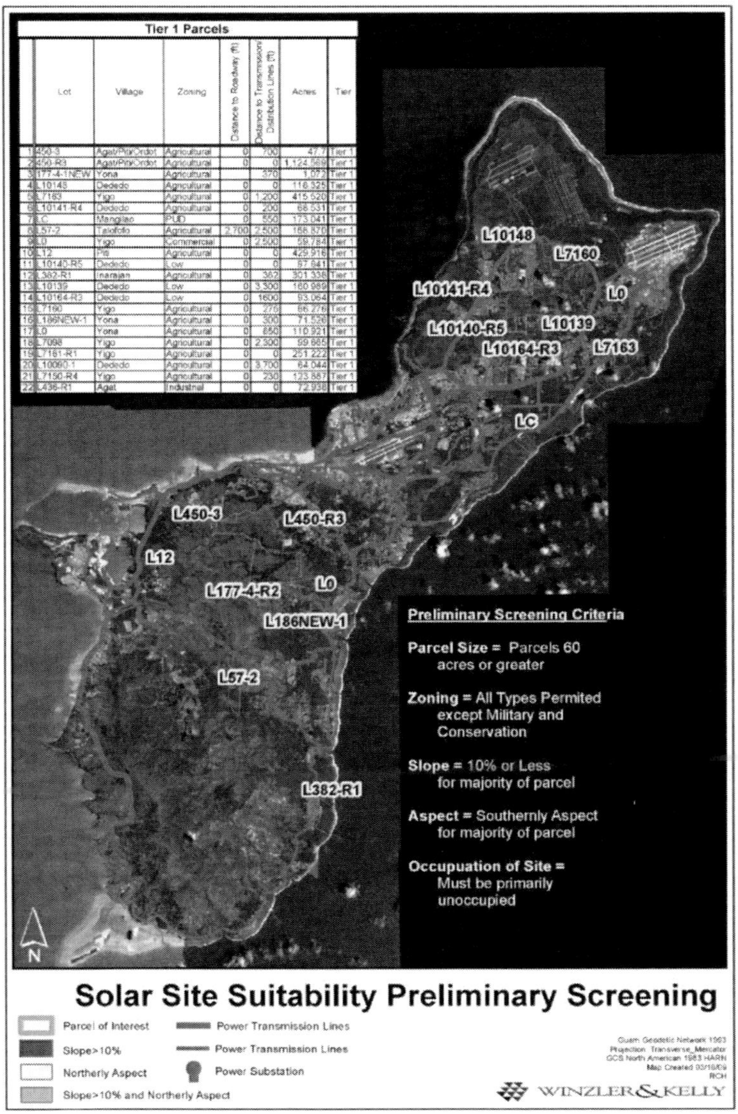

Source: GPA.

Figure 35. Overview of Tier 1 parcels.

All parcels were evaluated to identify contiguous areas internal to the parcel that were at least 60 acres and had the appropriate slope aspect. Slope aspect was not weighed as heavily as percent of slope.

Therefore, many parcels with the wrong slope aspect were permitted into Tiers 1 and 2, as long as the slope percent was acceptable.

For the Tier 1 parcels, distance to roadway was calculated and recorded.

For the Tier 1 parcels, distance to the nearest power transmission or power distribution lines was calculated and recorded. For an overview of Tier 1 Parcels, see Figure 35.

C.1.3. Top Seven Parcels

The top seven parcels from Tier 1 identified during this study are shown in Table 17. The third column is an estimate of the total amount of the array output at a conservative (relatively low) value for the array density, and the fourth column is an estimate at an optimistic (relatively high) value. Assuming every acre of every lot could be populated with PV modules, the potential for solar PV could range from 315 MW to 631 MW. What remains to be done now is to do field assessments of each site to determine how many acres of each site can actually accept solar arrays.

Table 17. Top Seven Sites Identified for Solar PV

Lot Number(s)	Total Size	100-kW / acre	200-kW / acre
	acres	kW PV	kW PV
L450-R3	1,125	112,450	224,900
L177-4-R2	1,072	107,200	214,400
L7163	416	41,550	83,100
L10140-R5 and L10164-R3	191	19,070	38,140
L57-2	169	16,890	33,780
L0	111	11,090	22,180
L186NEW-1	72	7,150	14,300
Totals	3,154	315,400	630,800

The remainder of this section briefly describes these seven parcels.

Parcel L57-2

This parcel is located roughly in the center of the South Island.

Advantages

- Large area with gradual slope and favorable aspect

- The land is largely undeveloped
- Reasonably close proximity to:
    - Electrical transmission/distribution lines
    - Access road
    - Electrical loads

Disadvantages

- Southern portion of parcel appears to have varied topography, including some areas with slope greater than 10% and northerly aspect
- Parcel appears to be heavily forested
- Not as close to electrical infrastructure, roads, and electrical loads as L7163, LO, or L186NEW-1

Parcel L177-4-R2

This parcel is located in the central part of the island.

Advantages

- Very large parcel
- The land is largely undeveloped
- Reasonably close proximity to:
    - Electrical transmission/distribution lines
    - Access road
    - Electrical loads

Disadvantages

- Substantial portions of parcel appears to have varied topography, including some areas with slope greater than 10% and northerly aspect

Parcel L186NEW-1

This parcel is located at roughly mid latitude on the island close to the Eastern shore in close proximity to Parcel LO, which is described above.

Advantages

- Close proximity to Parcel LO
- Parcels could be combined into one large project site
- Close proximity to:
- Electrical transmission/distribution lines

- Access road
- Electrical loads *Disadvantages*
- Parcel appears to have varied topography, including some areas with slope greater than 10% and northerly aspect
- Some structures do exist along the southern edge of the parcel
- Parcel appears to be heavily forested

Parcel L450-R3

This parcel is located in the central part of the island.

Advantages

- Very large parcel
- The land is largely undeveloped
- Near the Ordot Landfill
- Reasonably close proximity to:
    - Electrical transmission/distribution lines
    - Access road
    - Electrical loads

Disadvantages

- Substantial portions of parcel appears to have varied topography, including some areas with slope greater than 10% and northerly aspect

Parcel L7163

This parcel is located on a bluff on the North Eastern shore of the Island.

Advantages

- Large area with a high percentage of flat or gradually sloped land
- Close proximity to:
    - Electrical transmission/distribution lines
    - Access road
    - Electrical loads
- A significant portion of the property appears to have been cleared previously
- The land is undeveloped (i.e. no structures exist on parcel)

Disadvantages

- Potential impacts to recreation

Parcel LO

This parcel is located at roughly mid latitude on the island close to the Eastern shore.

Advantages

- Large area with a high percentage of flat or gradually sloped land
- Close proximity to:
  - Electrical transmission/distribution lines
  - Access road
  - Electrical loads
- A significant portion of the property appears to have been cleared previously
- The land is undeveloped (i.e. no structures exist on parcel)

Disadvantages

- None apparent

Parcels L10164-R3 & L10140-R5

These adjacent parcels are located roughly in the center of the North Island.

Advantages

- Together, these parcels have a reasonably large area with gradual slope and favorable aspect
- Close proximity to:
  - Electrical transmission/distribution lines
  - Access road
  - Electrical loads

Disadvantages

- Parcels both have some residential and agricultural development on them
- Parcel L10164-R3 does have some area that is greater than 10% slope.
  - However, this area is southerly facing.
- The undeveloped area on these parcels appears to be moderately forested
- Closest proximity to electrical infrastructure appears to be to either transmission or distribution lines, not both.
- Data was not sufficient to make a further determination on this point

C.1.4. Summary

Based on this preliminary study it appears that there are numerous properties available on Guam that could support a 10 MW solar power plant. However, potential developers must include provisions to perform a System Impact Study for any potential site they may consider. This should include a review of the adjacent GPA transmission and/or distribution systems, load centers, line losses, proximity to roads and improvements that may be required to satisfy other aspects of the IFB.

C.2. Wind Technology

C.2.1. Previous Wind Assessment Activities

Both GPA and the U.S. Navy have conducted wind development assessments on Guam. In a study conducted by DNV-Global Energy Concepts for GPA, two viable wind sites have been identified; the larger one is in the area of Cotal, along the Cross Island Road (HWY 17). The second site was in the area of Pulantat.

The Cotal area could support approximately 10 large wind turbines along two ridges, with a total capacity of approximately 20 MW. The site has good road access, is close to a 34.5 kV transmission line, and is within several miles of a substation. The site was reviewed during the NREL island visit and proved to be a good location. The turbines would be installed upwind from several homes and a larger potential development area. The site sits directly south of the Leo Palace Resort and although not close, will be clearly visible from the resort. The elevation of the site is around 200 m, which puts it only approximately 50 m lower that the island high spots and approximately 3 mi. from the U.S. Navy met mast, which shows an annual wind speed in the area of 8 m/s. Land in this area is primarily owned by the Department of Agriculture with several smaller private holdings. GPA has been working to implement a 50 m met tower at this site to allow detailed wind resource characterization and the development of project economics. The initial attempt to compete a RFP for the installation was hampered by the requirement of any measurement equipment being able to withstand a Class V typhoon. Due to the very high resulting costs proposed, the RFPs had to be canceled and next steps assessed.

The Pulantat site sits very close to the Leo Palace Resort and could support approximately five wind turbines with a total capacity of around 10 MW. This site is much more complicated due to proximity to local residents and facilities, local topography, the airport, and other communications/radar facilities. The site also sits on private land holdings and is lower in elevation then the Cotal site. This site is considered secondary to the Cotal location.

GPA has also conducted a long term wind measurement program as part of its air abatement activities. The met tower is a nominal 60 m tower and installed near the Cabras-Pili power plant. The site is generally sheltered and is on the leeward side of the island, so likely will not provide a good correlation for wind project assessment.

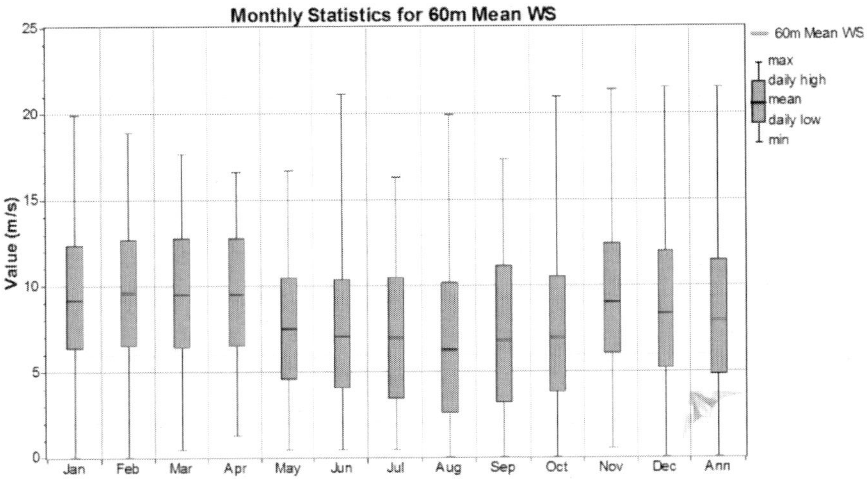

Source: NREL.

Figure 36. Boxplot of mean wind speed at 60 m for Guam Naval Ordnance (10/1/2008 – 9/30/2010).

The U.S. Navy, through an Interagency Agreement between the Naval Facilities Engineering Service Center (NFESC) and NREL, has been funding a wind resource assessment on U.S. Navy lands. As part of this project one 60 m met tower was installed on the western boundary of the Ordnance Annex just below Mt. Alifan. This measurement station was installed in September 2008 and although it would be very difficult, if not impossible, to install wind turbines in this area due to environmental impacts, it does provide a very good free wind speed reference site for any additional wind assessment programs.

The data collected thus far shows it to be a very good Class 5 resource with a mean wind speed in the region of 7.9 m/s. A monthly plot of wind speed is shown in Figure 36. The winds peak midday and come primarily from the east and south by southwest, as shown in Figure 37. Since the installation of turbines along the ridge is not likely in the near term, the U.S. Navy has considered installing turbines at other locations, such as the U.S. Navy port breakwater, for community-scale wind projects of one to three multi-MW wind turbines. The location and assessment status of these projects is not currently known but do represent additional installed wind capacity on the island.

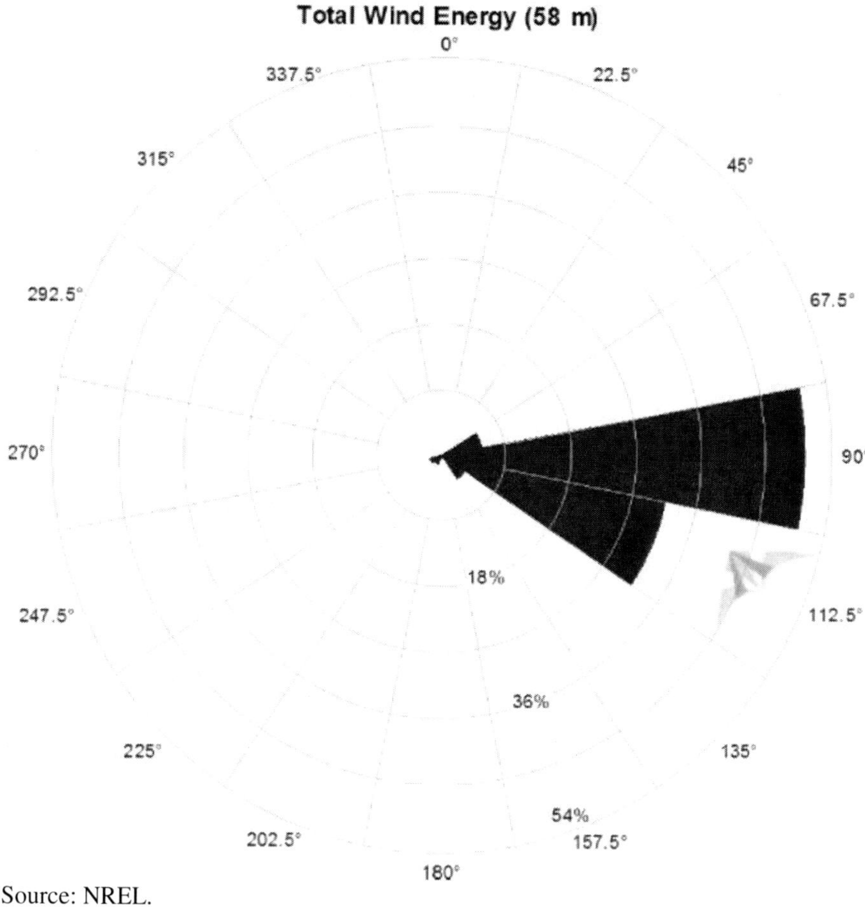

Source: NREL.

Figure 37. Power density rose for Guam Naval Ordnance (10/1/2008 – 9/30/2010).

Table 18. Estimated Total Population Increase on Guam from Off-Island (Direct, Indirect, and Induced)

	2010	2011	2012	2013	2014	2015	2016	2017	2018	2019	2020
Direct DoD Population[1]											
Active Duty Marine Corps	510	1,570	1,570	1,570	10,552	10,552	10,552	10,552	10,552	10,552	10,552
Marine Corps Dependents	537	1,231	1,231	1,231	9,000	9,000	9,000	9,000	9,000	9,000	9,000
Active Duty Navy[2]	0	0	0	0	0	0	0	0	0	0	0
Navy Dependents	0	0	0	0	0	0	0	0	0	0	0
Active Duty Army	0	50	50	50	50	630	630	630	630	630	630
Army Dependents	0	0	0	0	0	950	950	950	950	950	950
Civilian Military Workers	102	244	244	244	1,720	1,836	1,836	1,836	1,836	1,836	1,836
Civilian Military Worker Dependents	97	232	232	232	1,634	1,745	1,745	1,745	1,745	1,745	1,745
Off-Island Construction Workers (DoD Projects)[3]	3,238	8,202	14,217	17,834	18,374	12,140	3,785	0	0	0	0
Dependents of Off-Island Construction Workers (DoD Projects)	1,162	2,583	3,800	3,964	4,721	2,832	1,047	0	0	0	0
Direct DoD Subtotal	**5,646**	**14,112**	**21,344**	**25,125**	**46,052**	**39,685**	**29,545**	**24,713**	**24,713**	**24,713**	**24,713**
Indirect and Induced Population											
Off-Island Workers for Indirect/Induced Jobs[3]	2,766	7,038	11,773	14,077	16,988	12,940	6,346	4,346	4,346	4,482	4,482
Dependents of Off-Island Workers for Indirect/Induced Jobs	2,627	6,685	11,184	13,373	16,138	12,293	6,028	4,372	4,372	4,413	4,413
Indirect/Induced Subtotal	**5,393**	**13,723**	**22,957**	**27,450**	**33,126**	**25,233**	**12,374**	**8,718**	**8,718**	**8,895**	**8,895**
Total Population	**11,038**	**27,835**	**44,301**	**52,575**	**79,178**	**64,918**	**41,919**	**33,431**	**33,431**	**33,608**	**33,608**

Notes: [1] DoD population includes military personnel, DoD civilian workers and dependents from off-island.
[2] The Navy rows do not include increases from the transient presence of aircraft carrier crew with its CSG.
[3] Population figures do not include Guam residents who obtain employment as a result of the proposed actions.

Source: Guam and CNMI Military Relocation Final EIS (July 2010), Pg. ES-.8.

End Notes

[1] http://coris.noaa.gov/about/eco_essays/guam/geography.html.

[2] Government of Guam website: http://www.spc.int/prism/country/gu/stats/Census/Census.htm.

[3] CIA World Factbook: https://www.cia.gov/library/publications/the-world-factbook/geos/gq.html.

[4] OIA website: http://www.doi.gov/oia/Islandpages/gumpage.htm#pagetop.

[5] OIA website: http://www.doi.gov/oia/Islandpages/gumpage.htm#pagetop.

[6] Ibid.

[7] Weather Base, Guam: http://www.weatherbase.com/weather/weather.php3?s=021219&refer. Accessed January 7, 2011.

[8] Internal data provided to NREL by GPA.

[9] Source: Guam Energy Office. Spreadsheet provided by David Meno based on retail sales by fuel distributors (Shell, Mobil and 76).

[10] GPA March 2011.

[11] GPA March 2011.

[12] GPA March 2011.

[13] E-mail correspondence with Kenneth Kelly – NREL, March 2011.
[14] E-mail correspondence with Adam Warren – NREL, March 2011.
[15] GPA March 2011.
[16] GPA March 2011.
[17] GPA March 2011.
[18] CIA World Factbook: https://www.cia.gov/library/publications/the-world-factbook/geos/gq.html.
[19] By comparison, the U.S. Virgin Islands Water and Power Authority set a LEAC factor of $0.22/kWh in 2010.
[20] Secretariat of the Pacific Community: http://www.spc.int/prism/country/gu/stats/Utilities/Utilities.htm#PH5.
[21] GPA March 2011.
[22] Ibid.
[23] Energy Information Administration, http://www.eia.doe.gov/energyexplained/index.cfm?page=about_energy_efficiency.
[24] Guam Power Authority, Demand Side Management Plan, April 1993, Section II.
[25] End use data provided by GPA.
[26] GPA, Demand Side Management Plan, April 1993, Section II. This is the latest year that has data available.
[27] Data provided by the Guam Energy Office for November 2010.
[28] http://www.guamlegislature.com/Public_Laws_27th/P.L.%2027-132.pdf.
[29] Net Metering Program Interconnection Policy, 2/26/2009.
[30] http://www.guamenergy.com/announcement-to-appliance-retailers/.
[31] http://www.dsireusa.org/documents/Incentives/GU01R.htm.
[32] http://bcap-ocean.org/state-country/guam.
[33] Government of Guam Bill No. 460-30 (COR), passed October 28, 2010.
[34] http://www.guamlegislature.com/Public_Laws_29th/P.L.%2029-62%20%28Bill%20No.%20166%20EC%29.pdf.
[35] http://www.dsireusa.org/incentives/incentive.cfm?Incentive_Code=VI06R&re=1&ee=1.
[36] Ibid.
[37] Consent Decree for the Closure of Ordot Dump, Feb 2004, available online at http://iconguam.com/guamsolidwaste/docs/Federal%20Consent%20Decree%20[2-12-04%20in%20Civil%20Case%2002- 22].pdf accessed January 31, 2011.
[38] The Court Order can be found at http://www.guamsolidwastereceiver.org/pdf/court_order.pdf accessed January 31, 2011.
[39] *Quarterly Report of the Receiver*, August 11, 2010. pg 29. Accessed online on March 3, 2011: http://www.guamsolidwastereceiver.org/pdf/Tab_1_Quarterly_Report_of the_Receiver_8_11_10.pdf.
[40] *Quarterly Report of the Receiver*, December 9 2010, pg 34. Accessed online on March 3, 2011: http://www.guamsolidwastereceiver.org/pdf/Tab_1_Quarterly_Report_of_the_Receiver_12_9_10_Final.pd.
[41] *Estimation of Potential Yields of Landfill Gas from the Ordot Dump*. Duenas & Associates and URS Corp. July 2005. Accessed online at: http://guamsolidwastereceiver.org/pdf/RFP%20SWD001-11%20-%201b%20-%20Est.%20of%20Poten.%20LFG%20Yields%20Ordot%20Dump.pdf.. Accessed January 29, 2011.
[42] Solid Waste data provided by Captain Peter Lynch to Scott Haase via email March 20, 2011. Table is entitled Table 2.4-1 Projected Solid Waste Estimates, pg 2-109 and 2-110. All data

in the table are based on per capita waste generation estimates of 7.4 pounds per person per day.

[43] *Quarterly Report of the Receiver*, December 9 2010. Available on line at: http://www.guamsolidwastereceiver.org/pdf/Tab_1_Quarterly_Report_of_the_Receiver_12_9_10_Final.pd Accessed on March 3, 2011.

[44] Guam Solid Waste Weight Composition and Recycling Feasibility Study Final Report. Guam Environmental Protection Agency, May 19, 1995.

[45] Rossi Nayve Consulting Services and Associates, A Study of Guam's Recyclable Wastes, Guam Solid Waste Management Plan, 1982. Referenced online: http://www.co.hawaii.hi.us/env_mng/iswmp_orig/Chapter%2003.pdf, pg 5. Accessed February 1 2011.

[46] *Guam and CNMI Military Relocation, Final EIS*. Accessed online on March 3, 2011: http://www.guambuildupeis.us/documents/final/volume_6/Vol_06_Ch18_Hazardous_Materials_and_Waste.pdf.

[47] PL 25-175 is available online, http://www.guampowerauthority.com/special/documents/PL25-175_prohibitIncinerator_plandocs.pdf, accessed on 2/1/2011.

[48] DOE, *Guidelines for Selecting Cool Roofs*, July 2010: http://www1.eere.energy.gov/femp/pdfs/coolroofguide.pdf.

[49] http://www1.eere.energy.gov/femp/features/cool_roof_resources.html.

[50] http://www1.eere.energy.gov/femp/pdfs/fta_green_roofs.pdf.

[51] http://www1.eere.energy.gov/femp/features/cool_roof_resources.html.

[52] http://www.nrel.gov/docs/gen/fy04/36831m.pdf.

[53] http://www.nrel.gov/docs/fy99osti/26579.pdf.

[54] Walker, A., Christensen C., and Yanagi, G., "Time-of-Use Monitoring of US Coast Guard Residential Water Heaters with and without Solar Water Heating in Honolulu, HI" ASME International Solar Energy Conference 2003, March 2003, Kohala Coast, Hawaii.

[55] HECO Press Release. Available online at: http://www.heco.com/portal/site/heco/menuitem.508576f78baa14340b4c0610c510b1ca/?vgnextoid=9fbfb7ba04ecd210VgnVCM1000005c011bacRCRD&vgnextfmt=default&cpsextcurrchannel=1. Accessed 1/31/2011.

[56] Dr. Mari Marutani of the University of Guam conducted a survey of reasturants. Her data indicate approximately 3500 gallons/month. However, this survey did not capture all of the sources of grease. There are several grease haulers and pumpers on Guam and they indicated that the value of 3500 gallons/month is low compared to the actual volumes.

[57] *Estimation of Potential Yields of Landfill Gas from the Ordot Dump*. Duenas & Associates and URS Corp. July 2005. Accessed online at: http://guamsolidwastereceiver.org/pdf/RFP%20SWD001-11%20-%201b%20-%20Est.%20of%20Poten.%20LFG%20Yields%20Ordot%20Dump.pdf accessed January 29, 2011.

[58] *Opportunities for and Benefits of Combined Heat and Power at Wastewater Treatment Facilities*; Eastern Research Group and Energy and Environmental Analysis, U.S. Environmental Protection Agency: Washington, DC, 2007; Available online at: http://www.epa.gov/CHP/documents/wwtf_opportunities.pdf (accessed on March 28, 2011).

[59] EIA Guam Energy Profile.

[60] DOE Recovery Act State Memo, Guam (June 1, 2010): http://www.energy.gov/recovery/documents/Recovery_Act_Memo_Guam.pdf.

[61] ASHRAE Standard 55 – Thermal Environmental Conditions for Human Occupancy.

[62] US DOE-FEMP: Cool Roof Resources: http://www1.eere.energy.gov/femp/features/cool_roof_resources.html.

[63] http://www.rurdev.usda.gov/rhs/.

In: Securing Guam's Energy Future
Editor: Lucio N. Burrows
ISBN: 978-1-62948-397-9

Chapter 2

GUAM STRATEGIC ENERGY PLAN*

Misty Dawn Conrad

EXECUTIVE SUMMARY

This document presents various energy strategies that are available to Guam in the pursuit of diversifying fuel sources and reducing fossil energy consumption 20% by 2020 (20x20). The information in this document is based upon input from the Guam Energy Task Force (GETF), the National Renewable Energy Laboratory's (NREL's) *Guam Initial Technical Assessment Report*, and additional stakeholders. The information presented in this strategic energy plan is intended to be used by the GETF in developing an energy action plan. The next step in reaching the 20x20 goal involves selecting the most appropriate strategies from the options outlined in this strategic plan and creating a roadmap for implementation in the *Guam Energy Action Plan*.

The strategies are categorized based on the time required to implement them. Near-term strategies can be accomplished in 1-2 years, mid-term strategies require 3-5 years, long-term strategies require 6-8 years, and some strategies require 8 or more years before they can be fully implemented. The prioritization of the strategies will help facilitate the development of an action plan and the selection of appropriate strategies for meeting Guam's 20x20 goal. The strategies that will be

---

* This is an edited, reformatted and augmented version of a U.S. Department of the Interior Office of Insular Affairs publication, Produced by the National Renewable Energy Laboratory, dated July 2013.

used in reaching the 20x20 goal are broken down into five energy categories. The energy categories are shown in figure 1.

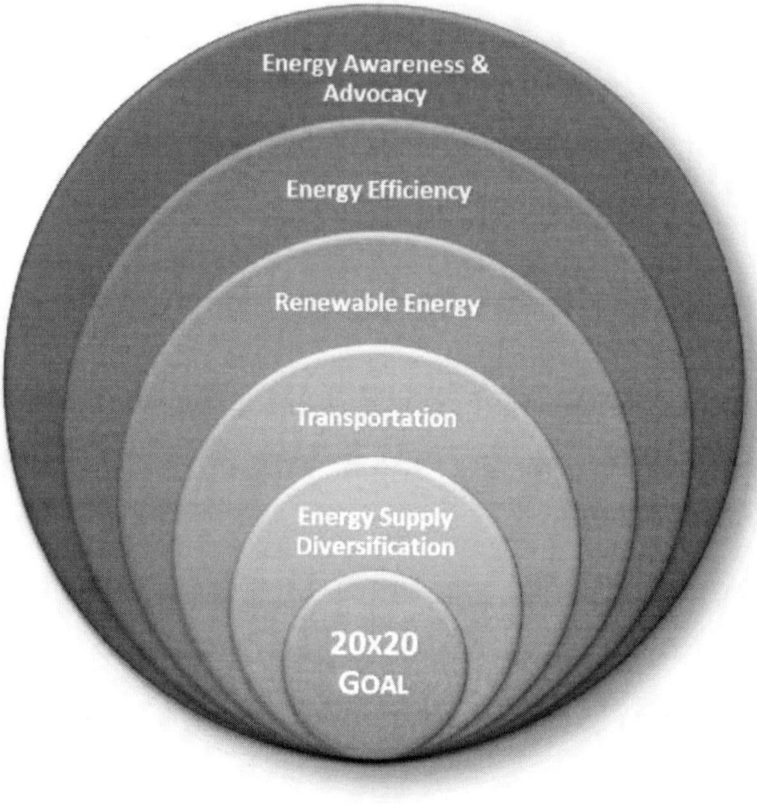

Figure 1. Areas to be addressed in order to reach Guam's 20x20 goal.

Energy education and outreach is addressed in detail in the report due to its impact on fossil fuel consumption in all of the energy categories. Effective energy education and outreach strategies can significantly reduce fossil fuel consumption and do not require large capital investments. Education and outreach strategies listed include building and energy code education and enforcement, providing energy efficiency courses for industry stakeholders, ensuring that energy projects remain transparent and accessible to the public, and using websites, events, and other outlets for renewable energy and energy efficiency education to the public. Strategies for implementing energy efficiency, renewable energy, transportation, and energy supply diversification are addressed and summarized in tables. Energy efficiency, renewable energy, and transportation strategy recommendations for Guam include:

- Use of energy conservation measures for buildings
- Implementation of a cool roof and cool pavement program
- Use of efficient outdoor lighting
- Increased efficiency in the utility's generation and distribution of electricity
- Use of waste-to-energy technologies
- Use of biofuels
- Use of wind energy
- Use of solar water heaters
- Use of solar photovoltaics
- Use of seawater air-conditioning
- Use of geothermal technologies
- Reductions in vehicle miles traveled
- Vehicle fuel-economy improvements
- Reductions in vehicle idle time
- Use of electric vehicles
- Use of biodiesel
- Improvements in traffic flow

Additionally, liquefied natural gas (LNG) is presented as a strategy for increasing Guam's energy supply diversification. The technology approach, technology details, strategy barriers, and time frame for each strategy are shown in the tables in the section titled Technology Options. Guidance for creating effective energy-related policies is provided, along with clean energy policy considerations for Guam. Additional information can be found in the Appendices and is intended to supplement the information found in the body of the document.

Introduction: The Urgent Need for Energy Transformation

Due to fluctuating energy prices, including the record high price for oil in 2008, and the anticipated military build-up, Guam has focused attention on the need to develop alternatives to its near 100% reliance on fossil fuels. Guam is in a unique position to undertake a concerted effort to address current energy and environmental challenges. Due to its geographic isolation, non-renewable resources are extremely limited and subject to large volatility in pricing and availability. Energy security is fundamental to Guam's economic future and sustainability. Creating a stable investment atmosphere remains challenging, while the quality of life for residents and visitors continues to be affected. However, the unique position of Guam also offers many natural advantages. The island is endowed with sources of alternative energy.

Guam's energy goal of reducing fossil fuel consumption 20% by the year 2020 encourages Guam to rise to the challenges created by volatile energy prices, increasingly scarce resources, vigorous world competition and technological innovation. Based on the principle of sustainability; Guam must meet the needs of today without compromising the ability of future generations to meet theirs. Variability on Guam, such as fuel pricing and the cost of energy, natural disasters, uncertainty about the extent of the military build-up, and fluctuations in income from tourism, make it difficult to achieve economic security. Sustainability is more than an environmental policy; it is an economic one as well. Energy touches every sector of a community, with far-reaching consequences when prices soar, stability is threatened or environmental impacts realized. Saving energy saves money and indigenous energy creates economic self-reliance. This new green economy approach is being implemented around the world. It is utilizing mature technologies, sparking innovation of new technologies, opportunities, jobs and economies.

By investing in green technology, energy conservation, energy efficiency, renewable energy, efficient transportation, green building design and smart grid technologies, Guam is building a new energy infrastructure that reduces its reliance on fossil-fueled centralized power and increases its use of green distributed power. Creating efficient systems that are integrated leads to security, resource efficiency, and sustainability. By investing in green business and technology Guam can be sustainable in the way it grows. A green infrastructure, powered by green technology, creates jobs.

By embracing the energy transformation process, Guam participates in this new economic framework by positioning itself at the forefront of change and innovation. Oftentimes, the most detrimental option is to continue to follow the status quo. While this is an easy route to take, the future energy security of Guam is dependent upon changes to its current approach.

The Energy Planning Process

In March of 2010, the U.S. Department of the Interior (DOI) Office of Insular Affairs (OIA) sponsored a regional energy meeting in Golden, Colorado, that included Guam, the CNMI, and American Samoa. Guam delegates met with representatives from the U.S. Department of Energy (DOE) and senior principals from DOE's National Renewable Energy Laboratory (NREL) to discuss ways to improve energy efficiency and increase the deployment of renewable energy technologies in the Pacific. As a result of this meeting, the Governor of Guam established the Guam Energy Task Force to help coordinate energy policy and promote long-term planning.

In the summer of 2010, OIA funded NREL to conduct an initial technical energy assessment for Guam[1] that detailed energy consumption and production data to establish an energy baseline. This assessment was used to conduct an energy analysis that estimated the energy efficiency and renewable energy potential for Guam.

The *Guam Initial Technical Assessment Report* was published in April 2011, and was used by the GETF as the starting point for developing this *Guam Strategic Energy Plan.*

The purpose of Guam's strategic energy plan is to provide project options to be further investigated and detailed in an implementation plan, or energy action plan. Developing an action plan is the next step in the energy transformation process. The *Guam Energy Action Plan* will select a handful of specific strategies that are achievable in the short-term (1-3 years), break down those strategies into feasible incremental steps, identify the stakeholders and desired outcomes, identify the organizations and individuals responsible for implementing the actions, and set a timeline for each step. The *Guam Strategic Energy Plan* will contribute to the development of the energy action plan by providing vital strategy options and information on those options.

Establishing the parameters of the strategic energy plan provides focus for the strategic planning team. Certain topic areas, such as water, are considered important to overall energy strategies, however were not considered within the Guam assessment due to budget constraints and the priority of targeting energy efficiency and renewable energy.

There are many energy efficiency and renewable energy technologies on the market today. Included within the *Guam Initial Technical Assessment Report*'s parameters were mature technologies that are commercially available and feasible. There are other technologies that Guam may want to investigate for future study, but for the basis of this report, technologies included have the following criteria:

- Commercially available
  - Tested and demonstrated
  - Carry warranty
- Service and parts available
- Ready for immediate deployment
- Economically viable
- Financing available from private sector organizations.

Technologies are constantly changing; therefore it is important to understand that assessments need to be updated to reflect these changes. Strategies for renewable energy and energy efficiency technology, modular nuclear, and LNG require further investigation to quantify the impact of specific technologies, programs, and/or projects. Environmental, regulatory, legislative, technical, and financial considerations will also need to be addressed during the energy planning and project development process.

THE GUAM ENERGY GOAL

Guam's goal is to secure its energy future and increase autonomy by diversifying fuel sources and *reducing fossil fuel energy consumption 20% by 2020 (20x20 goal).* This includes all energy consumed through power generation (utilities) and ground based transportation (vehicles), but does not include air and sea transportation. The GETF is responsible for facilitating Guam's efforts to reach the 20x20 goal.

Strategies for Achieving the Goal

The GETF has a multitude of approaches it can implement to achieve its 20x20 goal. These approaches, as shown in figure 2, are broken down into five categories: energy awareness and advocacy, energy efficiency, renewable energy, transportation, and energy supply diversification. Each approach contains strategies that can be utilized to successfully reduce fossil fuel consumption. To facilitate the success of these strategies they are broken down by the amount of time required to plan, assess, and implement them. On the path to achieving the 20x20 goal, the strategies will be implemented in three phases: near-term (1-2 years), mid-term (3-5 years), and long-term (6-8 years).

Additional projects that will not be completed by 2020 can be placed in a longer time frame. The time structure will allow the GETF to focus on a select group of strategies and set project completion milestones.

The GETF has a multitude of approaches it can implement to achieve its 20x20 goal. These approaches, as shown in figure 2, are broken down into five categories: energy awareness and advocacy, energy efficiency, renewable energy, transportation, and energy supply diversification. Each approach contains strategies that can be utilized to successfully reduce fossil fuel consumption. To facilitate the success of these strategies they are broken down by the amount of time required to plan, assess, and implement them. On the path to achieving the 20x20 goal, the strategies will be implemented in three phases: near-term (1-2 years), mid-term (3-5 years), and long-term (6-8 years). Additional projects that will not be completed by 2020 can be placed in a longer time frame. The time structure will allow the GETF to focus on a select group of strategies and set project completion milestones.

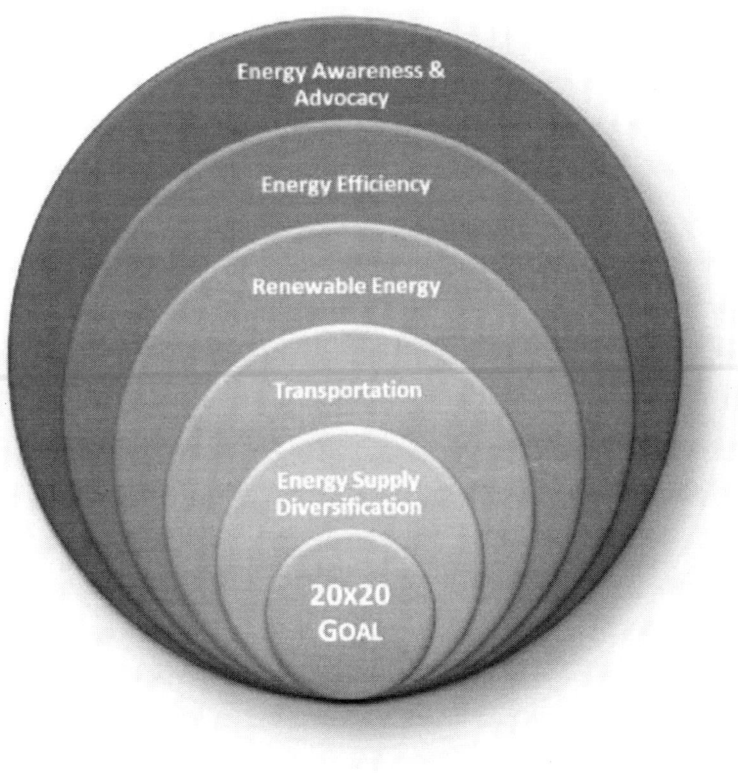

Figure 2. Areas to be addressed in order to reach Guam's 20x20 goal.

Metric for Evaluation

Progress towards the Energy Task Force's goal will need to be measured to determine if it has been met. All fossil fuel consumption will be compared to 2010 levels. Jet fuel reduction strategies are not being investigated and pursued as part of the reduction goal, and so jet fuel will not be included in the measurement calculation. The total number of gallons of fossil fuel consumed on Guam for any given year will be calculated by adding the total fuel sales by petroleum companies, minus jet fuel, to the amount of fossil fuels consumed by the Guam Power Authority (GPA). The number of gallons sold by petroleum companies is tracked by the Guam Energy Office[2], and GPA can provide the data on the annual amount of fossil fuel consumed for power generation. The total number of gallons of fossil fuels consumed will then be divided by the current population to provide per capita consumption. This measurement will account for growth. Census data is available for 2010, the baseline year, and will be available for 2020, the target year for achieving a 20% fossil fuel consumption reduction. For all other years during which census data is unavailable, population growth will be estimated based on GPA customer growth. The resulting measurement will then be compared to the identical measurement in 2010 to determine the reduction in fossil fuel consumption. Total fossil fuel use for Guam, not including jet fuel, in 2010 was 185,192,596 gallons and the population was 159,358, yielding an annual per capita fossil fuel consumption of 1,162 gallons.

GUAM ENERGY TASK FORCE (GETF)

The GETF was established by Executive Order 2010-15 on May 10, 2010. The GETF is a nonregulatory advisory group sponsored by the Government of Guam.

The GETF is comprised of local volunteers from multi-sector stakeholder communities and supported by federal agencies. Convened in September 2010, the GETF recruited appropriate leaders for membership to ensure diverse and equitable representation. The GETF was established to facilitate a transition towards a more sustainable energy future that will utilize the island's natural resources. The GETF is a forum for considering options and offering guidance related to the achievement of the island's energy goals. The task force's vision, mission, and goal are described below.

Vision

The GETF's vision is a secure, sustainable, and economically prosperous future for Guam.

Mission

As stated in Executive Order No. 2010-15, GETF's mission includes the following objectives:

- Reduction of reliance and expenditures on fossil fuels
- Development of indigenous and renewable energy sources
- Improvement of energy generation infrastructure
- Resource preservation, restoration and enhancement
- Training and education regarding energy efficiency and conservation
- Development of funding and financial strategies for sustainability and economic development
- Engagement in national and regional efforts to address island energy concerns
- Support energy literacy curriculum in all educational institutions
- Develop a vision, mission and goals for reducing dependence on fossil fuels, improving energy efficiency, and developing renewable energy resources
- Establish an organizational structure to appropriately develop a comprehensive, long-term strategic energy plan for Guam.

GETF Organizational Structure and Subcommittee Objectives

The GETF is led by an Executive Committee with four subcommittees: energy technologies, policy and finance, education and outreach, and federal facilities (figure 3). The four subcommittees and their focuses are described below.

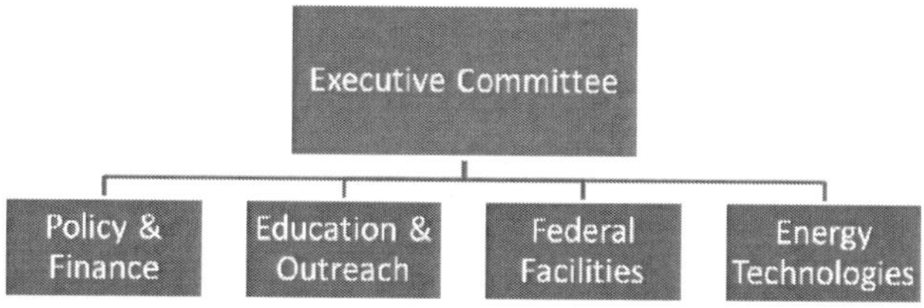

Figure 3. GETF organizational structure.

Policy and Finance

Governmental policy is foundational to reshaping the regulatory landscape to clear the way for a new energy future. The Policy and Finance subcommittee assists lawmakers and other clean energy advocates in analysis and development of policy opportunities to promote implementation of energy efficiency and renewable energy.

Education and Outreach

The Education and Outreach subcommittee is focused on identifying and developing educational tools and programs needed to develop and advance a clean energy economy. This included community outreach and education, technical and vocational training, and certification and degree programs.

Federal Facilities

In an effort to share information and facilitate dialogue, representation from the DOD has been essential. This subcommittee has been responding to federal mandates. They concentrate on energy conservation, building efficiency, outreach, renewable energy, and alternative vehicles.

Energy Technologies

The Energy Technologies subcommittee is focused on assessment and deployment of cost-effective energy efficiency and renewable energy opportunities. This includes energy conservation strategies, efficient building design, facility- and utility-scale renewable energy technologies, grid integration, and transportation.

POLICY FRAMEWORK AND PROJECT DEVELOPMENT

Policy is an effective tool in addressing clean energy adoption and subsequent development (figure 4), and the process of policy development addresses social concern through public participation. A policy is typically described as a principle or rule to guide decisions and achieve rational outcome(s).

The term is not normally used to denote what is actually done; this is normally referred to as either procedure or protocol.

Whereas a policy will describe the "what" and the "why," procedures and protocols address the "what," the "how," the "where," and the "when." Policies are generally adopted by the Board of, or senior governance body within, an organization whereas procedures or protocols would be developed and adopted by senior executive officers.

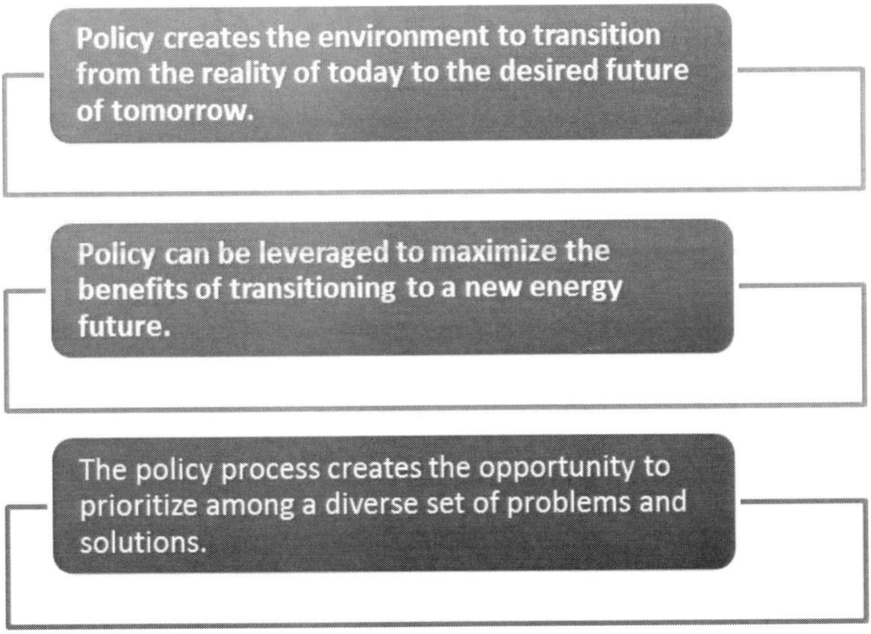

Figure 4. Advantages of a strong policy framework.

A precursor in directing Guam's energy transformation is to establish energy-related policies. The key principles that will guide Guam toward a prudent and sustainable energy future are listed below. The challenge in developing principles and policies is that at some level, inconsistencies, gaps and overlaps are inevitable — for example, when balancing sound economics with ecological integrity.

- Enhance the general quality of life through energy strategies that will benefit Guam
- Promote ecologically friendly propositions that create a healthy environment
- Promote environmental stewardship through energy conservation and energy efficient practice
- Support opportunities for local economic vitality
- Emphasize alternative power sources
- Support options that will reduce market volatility, stabilize rates, and increase reliability. It is important to have a robust set of policies that are:
- *Comprehensive* — Barriers to renewable energy and energy efficiency are diverse, growth may be halted from any number of wholly unrelated barriers
- *Broad-based* — Applies across all types of entities and sectors
- *Multifaceted* — Attacks problems and barriers from various angles
- *Multitiered* — Includes mandates or quotas but also addresses siting and permitting
- *Practical* — Focus on small victories
- *Long-term* — Create the market conditions to build a *real* and *sustainable* industry.

No single policy exists in isolation. Policy success is often determined by the ability to address multifaceted barriers with the appropriate mix of tools. One of the key benefits of policy development is that through this process, a specific set of challenges can be identified and strategies developed to address them; it does not necessarily attempt to be everything to everyone all at the same time. This process is incremental and topic-specific. A deliberate and methodical approach can help to minimize policy gaps and failures. Consider alternatives — think creatively.

Clean Energy Policy Considerations for Guam

• RENEWABLE PORTFOLIO STANDARD • ENERGY EFFICIENCY PORTFOLIO STANDARD • LEAD BY EXAMPLE • LATEST BUILDING CODES • ENERGY CODES • PERMITTING AND SITING STANDARDS √ NET METERING √ INTERCONNECTION	• APPLIANCE EXCHANGES AND REBATES • SMART METERING • TAX INCENTIVES • INDUSTRY RECRUITMENT INCENTIVES • REBATES AND GRANT PROGRAMS • LOAN AND LOAN GUARANTEES • PUBLIC BENEFIT FUND • PUBLIC FINANCING PROGRAMS

√ Denotes policy Guam has already implemented.

Policy Recommendations

The following are some policies that could attract greater investment in renewable energy and energy efficiency technologies and facilitate project development.

Develop Financial Mechanisms and Opportunities

1. Explore public-private partnerships with local businesses, developers, and banks with the idea of creating local markets for renewable power generation, energy efficiency, and energy conservation products and services.
2. Investigate third party financing
3. Pursue subsidies for renewable projects
4. Consider opportunities such as peak pricing, grants and subsidies to reduce the overall cost of electricity

Investigate alternative financing models such as power purchase agreements and performance savings' contracting.

Standardize the Development Process

In order to entice investment, communities are trying to reduce the level of project uncertainty as much as they can. Several strategies can be implemented, such as determining and offering quality information to potential investors through the request for proposal mechanism. This supplies potential investors with accurate information and the level of detail necessary to reduce their risk and increase their level of interest. This also casts a wider net, bringing in high quality companies. Providing information on a community's legal requirements for building a project — including site access, permitting and environmental processes, and what types of project ownership structures are recognized — can also be highly beneficial to potential investors. Communities are surveying their processes, eliminating unnecessary steps, and streamlining processes as an incentive to bring in investors. This strategy has multiple benefits — it organizes a community, encouraging people to survey what exists and develop what is needed; it provides consensus; and offers investors high quality information and reduces their uncertainty. A project development framework is provided in the Appendices.

Education and Outreach Recommendations

Education and Training

1. Code Enforcement

Once building and energy codes are updated, it will be necessary to educate the agencies and departments responsible for executing them and train enforcement officers for both inspection and enforcement capabilities.

2. Courses and Materials

Develop training courses and materials for both architects and workers in the construction sector to improve energy awareness. Energy efficiency in retrofit, renovation, and new construction can be included in design, materials, and equipment.

3. Guidebooks

Create guidebooks and self-assessment checklists. These guidebooks could be designed as standalone products for specific sectors or integrated into training courses described above.

Outreach Campaigns

1. ETF Visibility

It is important for the Energy Task Force to have visibility and credibility with all stakeholders, especially the citizens of Guam, so that its recommendations will have maximum impact. This can be accomplished by establishing a strong identity or brand for the ETF. One option is to embark on a contest within the school system for materials such as a logo, song, tag line, etc. Another option is to establish a website where all materials and calendars can be viewed and a blog or feedback loop implemented for people to comment on said materials and their concerns.

2. Energy Literacy

Create citizen buy-in in for new technologies, programs and policies by holding educational fairs, creating fact sheets, radio announcements/shows, electronic newsletters and a variety of activities that promote energy literacy from cradle to grave: understanding where energy comes from, how it is moved and used, how to make use more efficient, and how to conserve it and its life cycles.

Federal Facilities

It is essential to look at the energy picture of Guam holistically. The Department of Defense (DoD) has considerable representation on-island, accounting for between 21% and 22% of Guam's energy consumption. An extensive understanding of the relationship between Guam's DoD installations and the rest of the island is necessary to establish energy efficiency and renewable energy baselines and potential. NREL's assessment team concentrated on baseline factors taking into account DoD presence. This report focuses on island-wide utility generation and consumption patterns, not specific DoD energy efficiency and renewable energy activities. DoD has its own internal goals for renewable energy and energy efficiency. As DoD seeks to meet these goals, there may be impacts on the operations and economics of Guam Power Authority (GPA); it is important for DoD and GPA to continue to work together to understand these impacts and seek to develop win-win solutions.

Energy Technology Options

The GETF has numerous technology approaches it can implement to achieve its 20x20 goal. These technology approaches are broken down into four sectors: renewable energy, energy efficiency, transportation, and energy supply diversification. Each technology approach contains strategies that can be implemented to successfully reduce fossil fuel consumption. To facilitate the success of the strategies they are broken down by the amount of time required to plan, assess, and implement them. On the path to achieving the 20x20 goal (figure 5), the strategies can be implemented in three phases: near-term (1- 2 years), mid-term (3-5 years), and long-term (6-8 years). Additional projects that will not be completed by 2020 can be placed in a longer time frame. The time structure will allow the GETF to focus on a select group of strategies and set project completion milestones. Tables 1 through 4 present the technology options, technology details, barriers for implementing the technologies, strategies for overcoming the barriers, and the achievable time frames for implementation.

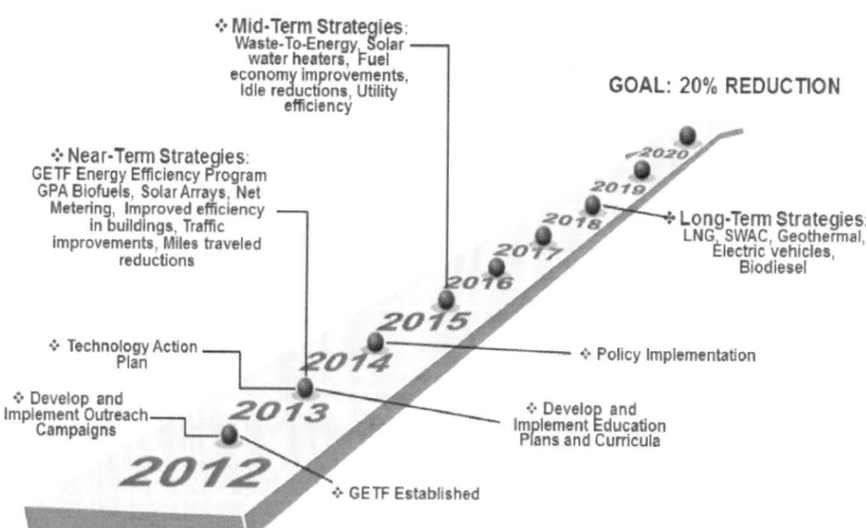

Figure 5. Timeline for achieving the energy goal.

Table 1. Technology Strategy: Energy Efficiency

N = NEAR-TERM M = MID-TERM
L = LONG-TERM B = BEYOND

Energy Efficiency				
Technology Type/Sector	**Technology Examples**	**Barriers**	**Strategies**	**Time Frame**
Buildings - Public (Existing)	• Measures identified in energy audits • Measures installed in EE retrofits • Additional energy conservation measures (ECM)	• Funding • Costs • Lack of on-site expertise and capacity to conduct energy audits	• Energy Audit Training Programs for on-site facility managers and staff	N
			• Energy assessments and modeling trainings	M
			• Revolving loan fund	L
Buildings - Residential (Existing)	• Low income subsidies • Accelerate weatherization programs • Efficient lighting, control settings, etc.	• Lack of "how to" knowledge • Lack of will to change behavior • Lack of knowledge of what is available to the consumer • Perception of high initial costs	• Awareness and education campaigns	N
			• Pilot a real-time feedback program utilizing smart measuring devices	L
			• Revolving loan fund	M
Buildings - Commercial (Existing)	• Energy audits • EE retrofits • Utilize ECMs	• Costs • Lack of on-site expertise and capacity to conduct energy audits	• Energy Audit Training Programs for on-site facility managers and staff	N
			• Energy assessment and modeling trainings	M
			• Revolving loan fund	M
Buildings - Industrial (Existing)	• Energy audits • EE retrofits • Utilize ECMs	• Costs • Lack of on-site expertise and capacity to conduct energy audits	• Energy Audit Training Programs for on-site facility managers and staff	N
			• Energy assessment and modeling trainings	M
			• Revolving loan fund	M
Buildings - New Construction & Major Renovations	• Adopt and enforce 2012 building codes • Integrated design and systems integration – HVAC, plumbing, electrical, lighting, etc. • Integrating renewables to achieve efficiency (e.g., building orientation)	Lack of: • Expertise and knowledge in employing integrated system design approach and whole building design practices • Expertise and capacity to operate and maintain such systems • How to measure bldg. performance	• Identify and publicize funding sources (ESPCs, etc.)	M
			• Provide informative resources describing whole building design practices and systems integration	N
			• Training programs for O&M and M&V managers and technicians	N
			• Revolving loan fund	M

Table 1. (Continued)

Energy Efficiency				
Technology Type/Sector	Technology Examples	Barriers	Strategies	Time Frame
Cool Roofs & Pavements	• Light colored surfaces (white or reflective)	• Potential increased short wave irradiation through large scale installation of reflective surfaces could increase cooling loads (unintended consequence).	• Develop a detailed land use database	M
			• Perform detailed analysis (Energy impacts)	M
			• Develop implementation programs (roofs, pavements, trees)	L
			• Develop a feedback system	M
			• Develop energy codes, standards, guidelines	N
			• Develop demonstration projects	M
Efficient Outdoor Lighting	• Fixture selection • Lighting standards • Energy consumption	• Initial costs and tight budgets • Concerns over lighting color • Proper design capacity	• Training on photometric analysis combined with life cycle cost and energy analysis	M
Utility Efficiency	• Transmission • Distribution • Generation	• Funding • Load balancing challenges • Voltage optimization	• Loss assessment study	N
			• Ensure that all consumption is metered	M
			• Waste heat recovery	L
			• System upgrades	M

Table 2. Technology Strategy: Renewable Energy

N = NEAR-TERM M = MID-TERM L = LONG-TERM B = BEYOND

Renewable Energy				
Technology Type	Technology Examples	Barriers	Strategies	Time Frame
Biomass/ Waste-To-Energy (WTE)	• Landfill gas (LFG) • Biomass • Anaerobic digestion	• Public Law PL 25-175[3] • Characteristics of the waste feed stock • Tipping fees are usually required to make WTE projects economically viable	• Form a working group to address WTE and PL 25-175	N
			• Repeal PL 25-175	M
			• Biomass resource assessment	M
			• WTE feasibility studies: Waste characterization, economic and technical analyses	M
Biofuel	• Ethanol • Coconut oil	• Limited land area • Limited feedstock supply	• Biofuel feasibility study	N

Renewable Energy				
Technology Type	**Technology Examples**	**Barriers**	**Strategies**	**Time Frame**
Wind	• Typhoon rated turbines • Turbines that can be lowered	• Cost • Integration • Social acceptance • Local capacity • Extreme weather events	• Erect MET towers for correlation to other data	N
			• Social acceptance and outreach program	N
			• Environmental and radar impact study	M
			• Investigate typhoon-rated turbines	M
			• Permitting for available sites	M
			• Environmental reviews, economic modeling, and interconnection studies	M
			• Wind farm project implementation	L
Solar Water Heaters (SWH)	• Flat-plate collector • Integral collector-storage systems • Evacuated-tube solar collectors	• Cost • Social acceptance of the technology	• Rebate program	M
			• Outreach program to encourage home energy and water conservation	N
			• Mandatory SWHs on government buildings	M
Solar PV	• Large-scale distributed systems • Off-grid systems • Must consider optimal panel tilt for efficiency and protection from extreme weather events • Must incorporate corrosion- and wind-resistant components	• Cost • Integration • Variability	• Proceed with Layon Landfill Solar Facility	N
			• Rebate for residences	M
			• Microclimate studies specific to planned PV sites	M
Sea Water Air Conditioning (SWAC)	• District cooling system	• Few commercial examples • High initial cost	• Feasibility study	M
			• Develop a business plan for implementation (permitting, financing, etc.)	L
			• Contract procurement	B
Geothermal	• Direct use geothermal • Electricity production • Heat pumps	• May not be cost-effective • Uncertainty over resource availability	• Assess geothermal potential	M
			• Environmental, economic, and interconnection studies	B

Table 3. Technology Strategy: Transportation

N = NEAR-TERM
M = MID-TERM
L = LONG-TERM
B = BEYOND

Transportation				
Technology Type/Goal	**Technology Examples**	**Barriers**	**Strategies**	**Time Frame**
Miles Traveled Reductions	• Public transportation system • Bicycle and pedestrian travel • Work from home and carpool programs	• Funding • Coordination between multiple sectors • Reliance on personal automobiles • Behavior change	• Rideshare website	N
			• Bike route designation	N
			• Education for alternative transportation	N
			• Coordination plan for taxis and GRTA	N
			• Bus tracking and coordination	N
			• Telecommute	N
			• Bike share system	N
			• Coordinate ADA compliance with biker and pedestrian friendly facilities	N
			• New buses and routes	N
Fuel Economy Improvements	• High miles per gallon vehicles • Efficient road system (reduce stopping events) • Hybrid vehicles	• Lack of knowledge of fuel saving practices • Inefficient vehicle fleet • Cost of fuel efficient vehicles	• Registration fee/rebate to encourage purchasing efficient vehicles	M
			• Education campaign for fuel economy	N
			• Motorcycle and bicycle safety plan	N
			• Mobile emissions testing	M
Idle Time Reductions	• Auxiliary power (Shorepower)	• Lack of knowledge of idling impacts • Difficulty in enforcing idling policies	• Idle reduction limitations for trucks and school buses	M
			• Auxiliary power (Shorepower) for passenger buses	M
Electric Vehicles	• Electric vehicles • Electric/solar-electric charging stations	• Cost • Fleet turnover rate	• EV trial fleets	L

Transportation				
Technology Type/Goal	Technology Examples	Barriers	Strategies	Time Frame
Biodiesel	• Waste grease and coconut oil esterification	• Collection and farming systems • Processing infrastructure	• Improved enforcement of grease disposal laws	N
			• Waste grease to biodiesel	M
			• Coconut oil to biodiesel	M
			• Construct a new biodiesel refinery for coconut oil processing	L
Traffic Flow Improvements	• Traffic signal coordination	• Funding	• Central coordination center	N

Table 4. Technology Strategy: Energy Supply Diversification

N = NEAR-TERM M = MID-TERM L = LONG-TERM B = BEYOND

Energy Supply Diversification				
Technology Type	Technology Examples	Barriers	Strategies	Time Frame
LNG (Will not contribute to fossil fuel reduction goal)	• Import terminal • LNG open rack vaporizers or submerged combustion vaporizers • Dual fuel generators	• Capital cost • Investment risks • Conversion of existing system	• RFP to acquire a bidder for land, permitting, engineering, commissioning, warranties, and O&M	M
			• Reduce risk: design build or design build operate approach; fixed price contract; contract with one lead contractor; convert existing units early to dual fuel ability	L

Conclusion: Shifting Paradigms

Guam is taking a bold step to increase energy security and strengthen its economy. In doing so, Guam seizes an opportunity to lead, charting a course for other islands to follow as they journey toward energy transformation.

Guam is envisioning ways to change how consumers make their purchasing decisions based on habit, cultural norms, and old ways of thinking. Creating new realities in which utilities make decisions based on a new green economy and "three generation thinking;" setting out on implementing policy strategies for legislators which will transcend policies originally based on political "experience" and party-line thinking and asking financiers to embrace the "Triple Bottom Line," instead of making investments based on traditional methods of risk assessment and analysis.

In order to evolve beyond the status quo:

- Consumers need to expand their knowledge of green energy technologies and make informed, sustainable choices
- Utilities will need to open the door to new ways of doing business, such as increasing efficiency, encouraging energy conservation, and interconnecting with distributed generators
- Governments will need to shift their focus from politics to the public interest by striving to reduce energy costs, developing sustainable energy policies, and embracing new, commercially proven technologies
- Investors will need to take advantage of incentives, such as RE subsidies, and take a long-term view when weighing the risks and returns of energy projects.

Most importantly, Guam is beginning to understand the criticality of energy security and embrace sustainability by actively supporting clean energy goals. Energy transformation involves a revolutionary shift in how entire communities *think* about and *use* energy.

Appendices

A.1. A Model for Energy Planning

The ETF subcommittees are participating in the community energy planning process described below and shown in figure 6.

Plan

The planning phase includes these steps:

- Identify and bring together stakeholders
- Develop a vision for the community's energy transformation.

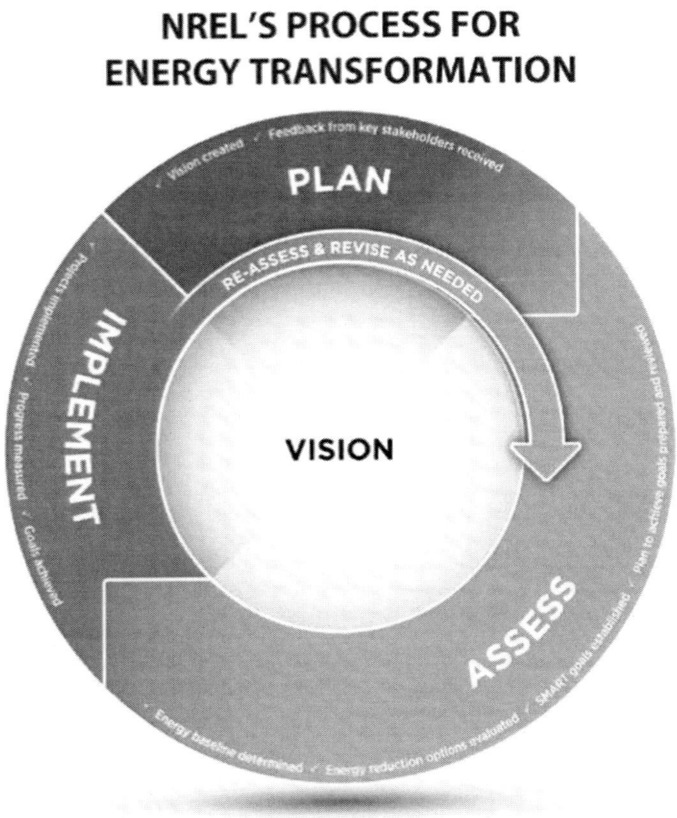

Figure 6. NREL's process for energy transformation.

Assess

The assessment phase includes these steps:

- Determine an energy baseline
- Evaluate options
- Develop goals
- Prepare an energy plan
- Solicit feedback on the plan from a broad set of stakeholders.

The goals of this phase are to:

1. Determine the community's energy baseline

2. Establish specific, measureable, attainable, relevant, and time-bound goals for energy transformation
3. Put the goals into a plan for action.

Implement

The implementation phase includes these steps:

- Develop, finance, and implement projects
- Create early successes
- Evaluate effectiveness and revise as needed. The goals of this phase are to:
  1. Implement projects that can build community support for ongoing and future energy projects
  2. Measure the progress and effectiveness of the plan and its projects.

A.2. Framework for Project Standardization

Reduce investor uncertainty by adopting a project development framework such as NREL's SROPTTC™[4] to examine the key issues and identify next steps towards resolving them. The following methodology outlines SROPTTC™:

- *Site* — Identify the physical location for the physical assets of a project, including property rights, length of tenure, terms and conditions, etc.
- *Resource* —Characterize and understand the renewable resource being considered.
- *Off-take* —Establish and secure by contract the buyer of both the energy and any other characteristics of output (e.g. renewable energy credits.
- *Permits* —Identify and obtain all permits necessary for project construction and operation.
- *Technology* —Invest in engineering design, equipment selection, and procurement activities of the chosen technology.
- *Team* —Assemble a fully qualified team that addresses all business, technical, financial, legal, and operational aspects.

- *Capital* —Attract financial resources necessary for final development, construction, commissioning, and initial operations.

A.3. Reduction Scenarios – Wedge Analysis

Methodologies and Assumptions

The wedge analysis was done using information provided from the Guam Power Authority and various stakeholders to establish a business as usual (BAU) case or base case. From this, and information gathered for the *Guam Initial Technical Assessment Report*, low- and high-impact scenarios were created reflecting 20% and 53% fossil fuel reduction and then compared to the base case. The wedge assumes a linear 0.119% growth in electricity consumption in the future, a rate that was calculated from utility data from the last ten years of consumption. The wedge analysis, while including some small specific renewable energy projects, is designed for high level analysis to depict potential fuel reductions from various commercially available and ready-to-deploy technologies. It assumes no major fluctuations in fuel prices by estimating the average fuel cost for the past ten years.

Base Case

Under the base case (figure 7), it is assumed that no significant steps are taken to curb fossil fuel use. It depicts the energy use in a business as usual scenario, and is used to compare the potential impacts of deploying energy efficiency measures and alternative technologies. The base case reflects the fact that Guam's electricity is generated mainly from fossil fuel.

Figure 7 illustrates the consumption of electricity in barrels oil equivalent (BOE) through 2026. This chart assumes 5,800,000 British thermal units (Btu's) for every barrel of oil. The unit of measure, barrels of oil equivalent, indicates how many barrels of oil would need to be consumed to meet the demand for electrical energy.

20% Reduction Scenario

Figure 8 illustrates one potential path to reducing fossil fuel consumption by 20%. The top of the graph represents the same linear increase as the base case. The red wedge shows the direct reduction of energy not needing to be produced due to potential efficiency improvements in generation and/or end-use consumption. The green wedge represents the potential contribution of various renewable energy technologies to the 20% reduction in fossil fuel consumption.

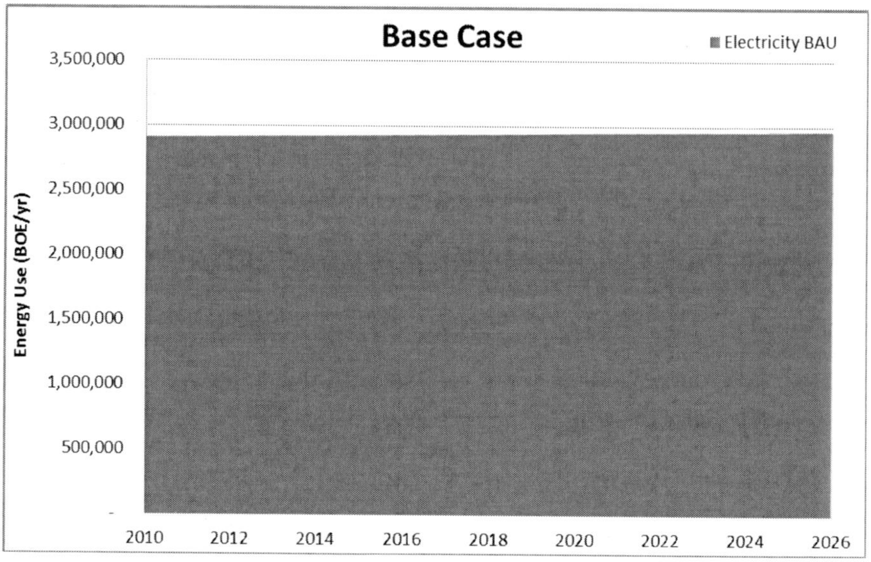

Figure 7. Base case.

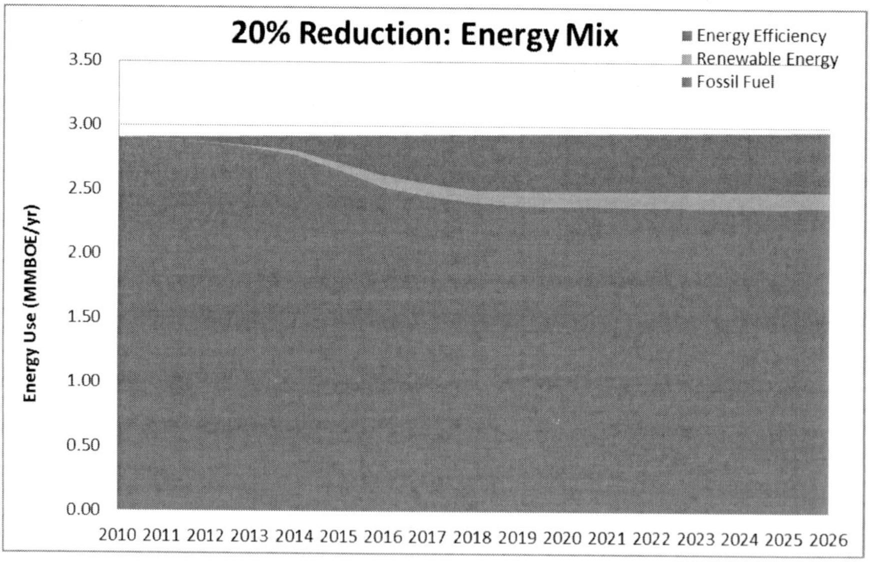

Figure 8. 20% reduction: energy mix.

Figure 9 represents the same information shown in figure 8 but in more detail. It shows how several renewable energy technologies, and energy efficiency actions in specific end-use sectors, could contribute to the overall reduction of fossil fuel consumption in Guam. Note that the legend is in the same order as each wedge is stacked. Similarly, the area cut out by energy efficiency improvements matches that of the red wedge in figure 8; additionally, the renewable energy wedge matches the same area in both charts.

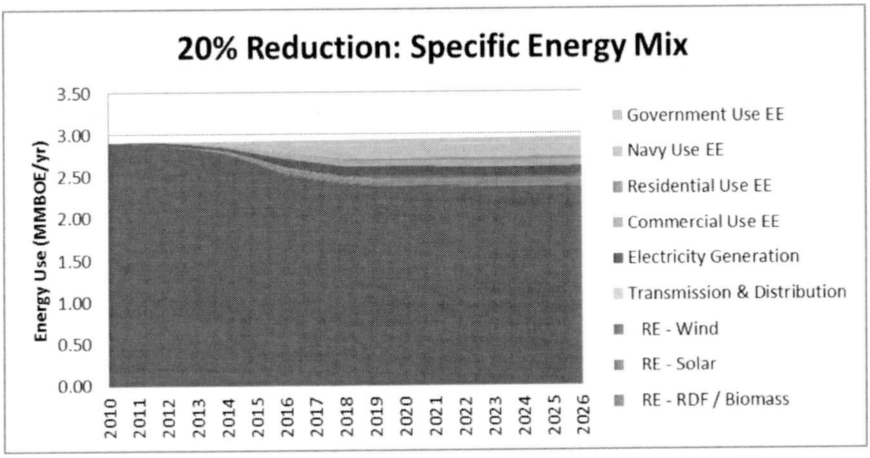

Figure 9. 20% reduction: specific energy mix.

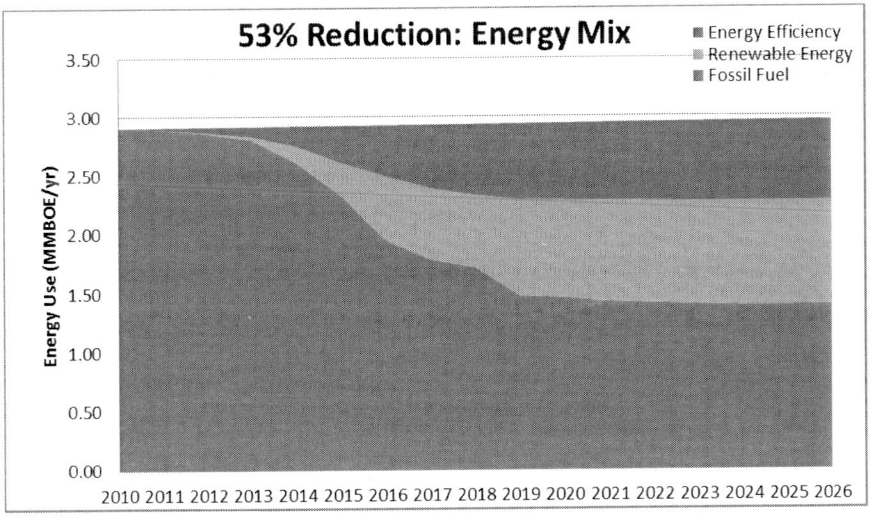

Figure 10. 53% reduction: energy mix.

53% Reduction Scenario

The high-impact scenario (figure 10) was determined by considering strategies that could meet a 53% fossil fuel reduction goal. The top of the graph again assumes the linear increase as the base case. The red wedge shows the direct reduction of energy not needing to be produced due to efficiency improvements in generation or end-use consumption while the green wedge represents the contribution of various renewable energy technologies to the reduction in fossil fuel consumption.

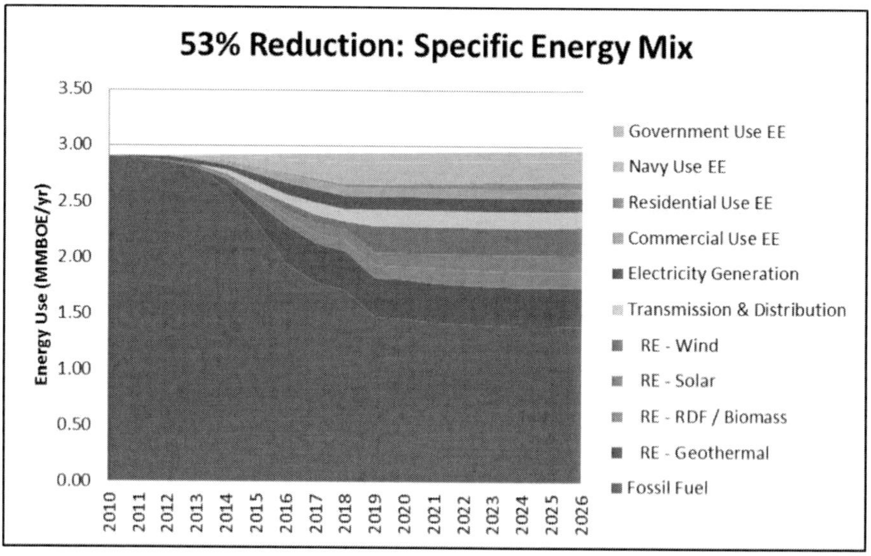

Figure 11. 53% reduction: specific energy mix.

Figure 11 represents the same information shown in figure 10 but in more detail. It shows how several renewable energy technologies, and energy efficiency actions in specific end-use sectors, would contribute to the overall reduction of the fossil fuel consumption in the energy portfolio for Guam. Note that the legend is in the same order as each wedge is stacked. Similarly, the area cut out by energy efficiency improvements matches that of the red wedge in the low impact energy mix chart above; additionally, the renewable energy wedge matches the same area in both charts.

Comparing all scenarios, one can see how implementing a wide range of energy efficiency and renewable energy technologies over time can reduce fossil fuel consumption and begin to paint a picture of sustainability. Strategy implementation takes time and so does project development. Change does not happen overnight, but with decisive goals and clear action steps, change will be incremental.

A.4. Proposed Transportation Projects

Proposed transportation projects[5] are listed in table 5, ranked first by cost (low to high), then by anticipated impact on petroleum reduction (high to low), then by popularity (high to low), then by speed of implementation (fast to slow).

Table 5. Transportation Projects by Rank

RANK	STRATEGY	PROJECT	COST	PETROLEUM REDUCTION	POPULARITY	TIME FRAME
1	Vehicle Miles Traveled (VMT)	Rideshare website	Low	High	High	Fast
2	Biodiesel	Waste grease to biodiesel	Low	High	Med.	Fast
3	Idle Reductions	Idle reduction initiatives for delivery trucks and school buses	Low	High	Med.	Med.
4	Fuel Economy	"Feebate" for improved fuel economy	Low	High	Low	Slow
5	VMT	Bike Route	Low	Med.	High	Med.
6	Fuel Economy	Education campaign for fuel economy	Low	Med.	Med.	Med.
7	VMT	Education campaign for alternative transportation	Low	Med.	Med.	Med.
8	VMT	Coordination plan for taxicabs and GRTA	Low	Med.	Med.	Med.
9	Fuel Economy	Guam motorcycle and bicycle safety strategic plan	Low	Low	Med.	Slow
10	VMT	Bus tracking and coordination system	Med.	High	High	Fast
11	VMT	Telework	Med.	High	High	Med.
12	Fuel Economy	Mobile emissions testing facility	Med.	Med.	Low	Med.
13	VMT	Bike share system	Med.	Low	High	Fast

Table 5. (Continued)

RANK	STRATEGY	PROJECT	COST	PETROLEUM REDUCTION	POPULARITY	TIME FRAME
14	Idle Reductions	ShorePower for buses	Med.	Low	High	Med.
15	VMT	Coordinate ADA compliance with bike-and pedestrian-friendly facilities	Med.	Low	Med.	Fast
16	Biodiesel	Coconut oil into B5	Med.	Low	Med.	Slow
17	VMT	New buses and routes	High	High	High	Fast
18	VMT	Port pipeline project	High	High	Low	Med.
19	Flow	Traffic signal coordination	High	Med.	High	Med.
20	Electric Vehicles	EV trial fleets	High	Med.	Med.	Med.

End Notes

[1] Ian Baring-Gould et al., Guam Initial Technical Assessment Report. National Renewable Energy Laboratory, NREL/TP-7A40-50580, April 2011.

[2] http://www.guamenergy.com/?s=petroleum

[3] http://guamlegislature.com/Public Laws 25th/P.L.%2025-175.pdf, accessed 6.27.2013

[4] http://www.nrel.gov/tech deployment/project development model.html

[5] Johnson, C. (2013). Guam Transportation Petroleum-Use Reduction Plan. NREL Report No. TP-7A30-57191.

In: Securing Guam's Energy Future
Editor: Lucio N. Burrows
ISBN: 978-1-62948-397-9

Chapter 3

GUAM ENERGY ACTION PLAN*

Misty Dawn Conrad and J. Erik Ness

DOCUMENT OVERVIEW

This document describes the four near-term strategies selected by the Guam Energy Task Force during action planning workshops conducted in March 2013, and outlines the steps being taken to implement those strategies.

Each strategy addresses one of the energy sectors identified in the Guam strategic energy plan as being an essential component of diversifying Guam's fuel sources and reducing fossil energy consumption 20% by 2020 (20x20) — see figure 1, below.

The four energy strategies selected are:

5. Expanding public outreach on energy efficiency and conservation
6. Establishing a demand-side management revolving loan program
7. Exploring waste-to-energy options
8. Influencing the transportation sector: anti-idling legislation, vehicle registration fees, and electric vehicles

---

* This is an edited, reformatted and augmented version of a U.S. Department of the Interior Office of Insular Affairs publication, Produced by the National Renewable Energy Laboratory, dated July 2013.

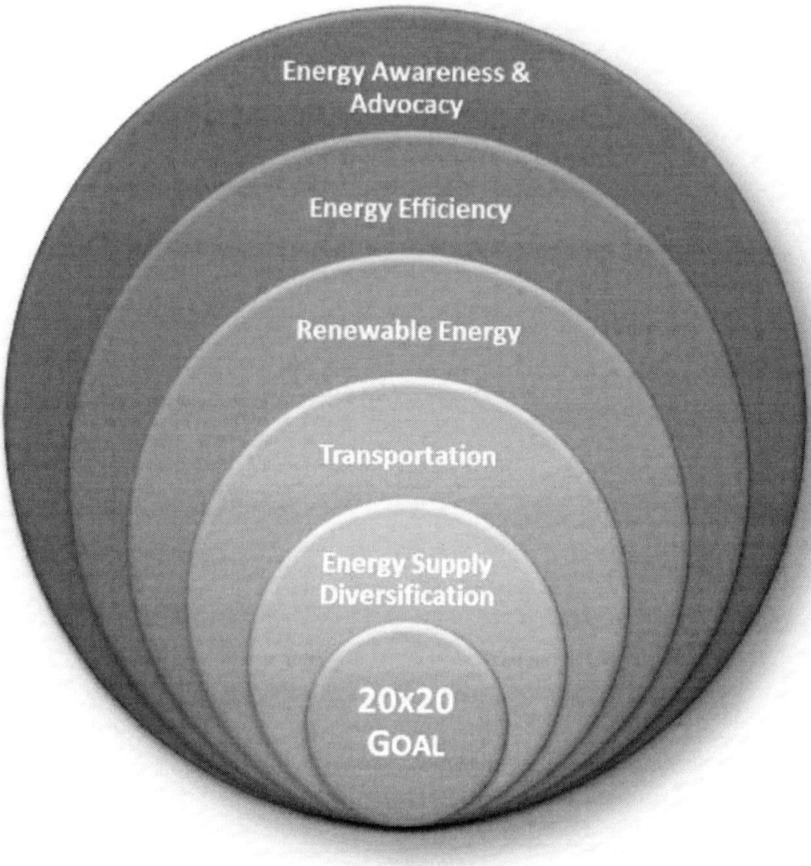

Figure 1. Options for reaching Guam's 20x20 Goal, as identified in the strategic energy plan.

GUAM ENERGY TASK FORCE

The Guam Energy Task Force (GETF) was established by Executive Order 2010-15 on May 10, 2010.

Vision

A secure, sustainable, and economically prosperous future for Guam.

Mission

As stated in Executive Order No. 2010-15, GETF's mission includes the following objectives:

- Reduction of reliance and expenditures on fossil fuels
- Development of indigenous and renewable energy sources
- Improvement of energy generation infrastructure
- Resource preservation, restoration and enhancement
- Training and education regarding energy efficiency and conservation
- Development of funding and financial strategies for sustainability and economic development
- Engagement in national and regional efforts to address island energy concerns
- Develop funding and financial strategies for sustainability and economic development
- Engage in national and regional efforts to address island energy concerns
- Support energy literacy curriculum in all educational institutions
- Develop a vision, mission and goals for reducing dependence on fossil fuels, improving energy efficiency, and developing renewable energy resources
- Establish an organizational structure to appropriately develop a comprehensive, long-term strategic energy plan for Guam

GETF Organizational Structure and Subcommittee Objectives

The GETF is led by an Executive Committee with four subcommittees: energy technologies, policy and finance, education and outreach, and federal facilities (figure 2). The GETF is a nonregulatory advisory group sponsored by the current administration. The GETF is comprised of local volunteers from multi-sector stakeholder communities and supported by federal agencies.

Convened in September 2010, the GETF recruited appropriate leaders for membership to ensure diverse and equitable representation. It facilitated a transition toward a more sustainable energy future that will utilize the island's natural resources. The GETF is a forum for considering options and offering guidance related to the achievement of the island's energy goals.

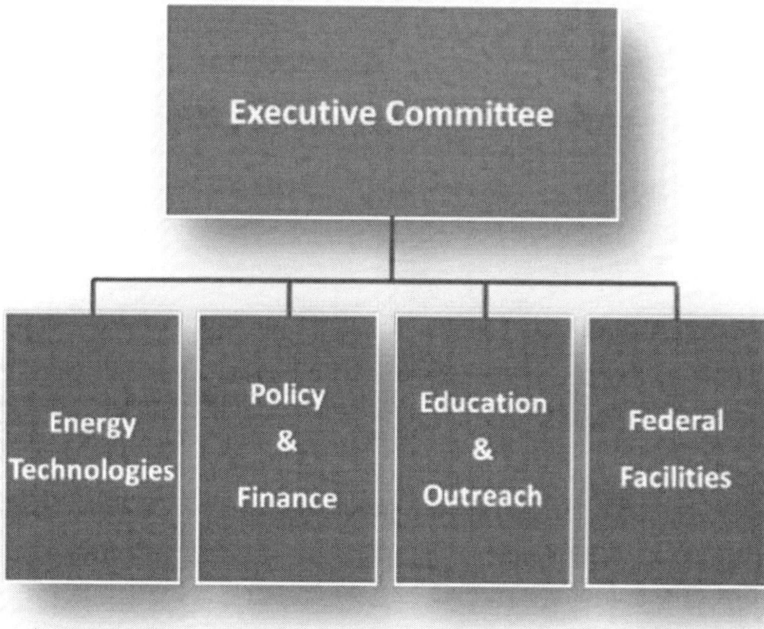

Figure 2. GETF organizational structure

The four subcommittees and their focuses are described below.

Energy Technologies

The Energy Technology Subcommittee is focused on assessment and deployment of cost-effective energy efficiency and renewable energy opportunities. This includes conservation strategies, efficient building design, facility- and utility-scale renewable energy technologies, grid integration, and transportation.

Policy and Finance

Governmental policy is foundational to reshaping the regulatory landscape to clear the way for a new energy future. The Policy and Finance Subcommittee assists lawmakers and other clean energy advocates in analysis and development of policy opportunities to promote implementation of energy efficiency and renewable energy.

Education and Outreach

The Education and Outreach Subcommittee is focused on identifying and developing educational tools and programs needed to develop and advance a clean energy economy. This includes formal and informal programs spanning community outreach and education, technical and vocational training, and certification and degree programs.

Federal Facilities

In an effort to share information and facilitate dialogue, representation from the DOD has been essential. This subcommittee has been in full swing responding to federal mandates. It concentrates on conservation, meter and building efficiency, outreach, and renewable energy and alternative vehicles.

THE ENERGY ACTION PLANNING PROCESS

In 2008, record high oil prices focused attention on the need to develop alternatives to Guam's near-100% reliance on imported fossil fuels. Due to its geographic isolation, nonrenewable resources are extremely limited and subject to large volatility in pricing and availability. Guam recognized that creating energy security is key to its future economic development and sustainability, and that proper long-term planning is required to achieve energy security.

In March of 2010, the U.S. Department of the Interior (DOI) Office of Insular Affairs (OIA) sponsored a regional energy meeting in Golden, Colorado, that included Guam, the CNMI, and American Samoa. Guam delegates met with representatives from the U.S. Department of Energy (DOE) and senior principals from DOE's National Renewable Energy Laboratory (NREL) to discuss ways to improve energy efficiency and increase the deployment of renewable energy technologies in the Pacific. As a result of this meeting, the Governor of Guam established an energy task force to help coordinate energy policy and promote long-term planning.

In the summer of 2010, OIA funded NREL to conduct an initial technical energy assessment for Guam[1] that detailed energy consumption and production data to establish an energy baseline. This assessment was used to conduct an energy analysis that estimated the energy efficiency and renewable energy potential for Guam.

The *Guam Initial Technical Assessment Report* was published in April 2011, and was used by the GETF as the starting point for developing the *Guam Strategic Energy Plan*, which was completed in 2012 and formally adopted in March 2013.

With continued funding from OIA and with the support of NREL staff, the GETF conducted action planning workshops in March 2013 to identify the near-term strategies that would be likely to have the greatest impact on reducing Guam's fossil-fuel energy consumption. Those strategies are described in detail in this document.

Unlike the strategic energy plan, which established the overall long-term goals for energy consumption and generation in Guam, this action plan selects a handful of specific strategies that are achievable in the short term (1-3 years), breaks those strategies down into feasible incremental steps, identifies the stakeholders and desired outcomes, identifies the organizations and individuals responsible for implementing the actions, and sets a timeline for each step.

When evaluating technology options, the GETF focused on mature, proven energy efficiency and renewable energy technologies that are commercially available and financially viable today.

Care was taken to ensure that the actions selected were not duplicating existing efforts, and that they leveraged work in other agencies and organizations as much as possible. The task force also gave considerable thought to ways of measuring the impact of the strategies in terms of reductions in fossil fuel consumption.

This action plan is a living document that will be regularly updated by the task force.

Programs and strategies will be modified or replaced as the task force gathers new information about different approaches to reducing energy consumption and as new technologies for generating electricity and reducing transportation fuel consumption emerge, or existing, but previously impractical, technologies become commercially viable.

During deliberations at several action planning workshops held in March 2013 and a follow-up discussion in April, group consensus was reached on the four strategies described in the following pages.

Strategy 1. Public Outreach on Energy Efficiency and Conservation

Energy awareness and outreach is one of the strategies from the strategic plan that was identified as having the greatest potential impact on energy consumption in Guam.

In order for the GETF's outreach campaigns to be have maximum impact, it is important for the GETF to be perceived by the citizens of Guam as a voice of authority. This involves raising awareness of the GETF and its role in tandem with energy education initiatives as this will maximize the impact of those initiatives.

A. GETF Branding

The GETF was envisioned as an independent organization that would carefully and methodically evaluate options for change, provide accurate information to the public, encourage changes in behavior, and recommend policy changes that could help ensure future energy security for Guam. Accomplishing these goals requires a hands-on approach, with the GETF helping to implement the solutions it recommends, working in collaboration with businesses, public sector agencies and the military.

The benefit of GETF branding is that, if the GETF has a strong identity as an independent authority, it can help to legitimize, strengthen and lend credibility to the energy outreach messages of other organizations, leveraging existing expenditures on energy and environmental education being carried out by other entities. It also benefits any smaller organizations that partner with the GETF, as they will get greater exposure through association with the GETF.

Existing organizations/initiatives that the GETF can partner with include:

- Guam Energy Office (GEO) — There is a direct relationship between recycling and energy use. GEO recently handed out recycling bins with this message on the side: "Recycling saves energy."
- Green Dream — Part of the Island-wide Beautification Task Force (IBTF), which is a very successful campaign.
- The i*recycle program.

- The University of Guam (UOG) Center for Island Sustainability (CIS) Energy Education.

The intention is not to be exclusive, rather for partner organizations to leverage each other's visibility in the eyes of the target audiences, promoting initiatives through joint messaging, such as "GETF in conjunction with the Green Dream ..."

Responsible for Coordinating Implementation: GEO and UOG's CIS

Potential Challenges

The branding itself is not a challenge. There is plenty of marketing talent on-island, including private companies and in-house designers and campaign staff at UOG. The only potential obstacle is covering the costs of developing the brand, which are not expected to be high.

Specific Actions and Timeline

1. Develop a work plan outlining the sequence of activities to accomplish this task, estimate the likely costs, suggest potential sources of funding and identify potential resources needed. The Education and Outreach Subcommittee will finalize the draft work plan and present it at the GETF meeting on June 14, 2013 for discussion and adoption. *[Education and Outreach Subcommittee, Carl Swanson*]
2. Secure funding (including first leveraging the resources of GETF members' organizations, such as UOG), and then issue an RFP for an existing marketing company to work with the GETF to implement the strategy if needed. *[Education and Outreach Subcommittee, Carl Swanson]*
3. Create a GETF website that provides information on the GETF and outreach campaigns. This could be started before the GETF has its own branding or campaigns, with the website acting as a portal with links to information about existing campaigns, GPA webpages describing rebates and incentives, etc. It would help establish the GETF as a source of information about energy in advance of its official "launch." The Guam Community College (GCC) may be able to help set up this website.

Reporting on Progress and Measuring Impact

Carl Swanson will report progress at the general GETF meeting scheduled for June 14, 2013.

B. Outreach Campaigns

Outreach campaigns will focus on promoting energy conservation (not using energy by doing without some service, such as turning the lights off) and energy efficiency (using less energy to accomplish the same task, such as using light bulbs that require less electricity) as it costs less to reduce energy consumption than it does to supply more energy using renewable or other technologies.

GEO is an important partner in this work as it has experience with implementing outreach campaigns, as is UOG, which has the existing skill set to measure the impact of those campaigns.

Before the GETF undertakes any outreach campaigns, it will identify existing campaigns and evaluate their effectiveness.

The GETF's intention is, as much as possible, to be a clearinghouse for information that others are producing, only creating its own information when it identifies a gap. Part of developing any new campaign will be to consider how to maximize the impact of the campaign using carefully targeted messaging to energy consumers and children, who in turn can influence their parents.

Possibilities for new information campaigns include:

- x Tips on energy conservation.
- x Helping people understand that efficiency improvements can mean spending less on their electric utility bills (which in turn means less power generation and less pollution). This could include promoting internet-based tools that can provide a cost-benefit analysis for the purchase of energy efficient technologies.
- x Public service announcements (PSAs) on how to get the most out of the smart meters being installed by the Guam Power Authority (GPA) .
- x Information on rebates for appliances and other energy-related investments (e.g., cool roofs) as they become available.

Responsible for Coordinating Implementation: GEO and UOG's CIS

Potential Challenges

Possible challenges include:

- x Securing funding and ensuring the long-term sustainability of funding to cover the time and expense involved in running information campaigns.
- x Ensuring consistency of messaging among organizations involved in energy campaigns: GETF, GPA, GEO, CIS, and others.
- x Ensuring that sources of information used by the GETF and partners are both reliable and "Pacific-centric" so that the messages are relevant to audiences in Guam.
- x Ensuring that outreach messages actually engage the average citizen.

Specific Actions and Timeline

1. Collect information on existing energy efficiency and conservation outreach initiatives. [*Peter Calvo with input from Peggy Denney and Heidi Ballendorf*]
2. Integrate Service Learning into the education strategy as a gateway to address K-12 education.
3. Identify metrics that could be used to measure the impact of the outreach initiatives, such as conducting surveys either on line or at events.
4. Develop a list of specific outreach messages and programs to be discussed and evaluated by the GETF. This will be done after information on existing initiatives is collected.
5. Create a GETF website that could act as a portal with links to information about existing campaigns, GPA webpages describing rebates and incentives, etc. This is the same website described under task 1.A.3, above.
6. Create opportunities for people to provide feedback and input on the outreach initiatives and ask questions. This could include an on-line blog, a cell phone app, and use of social media. Responsibility for this task will be determined after the GETF website is established.

Reporting on Progress and Measuring Impact

GEO and CIS will report progress at the general GETF meeting scheduled for June 14, 2013. Identifying suitable metrics is one of the action items listed above.

STRATEGY 2. DEMAND-SIDE MANAGEMENT REVOLVING LOAN PROGRAM

Energy-saving investments are very effective because people communicate their savings (from, for example, installing more-efficient air conditioners) to their neighbors and friends. Existing rebate programs have led to accelerated sales in efficient air conditioners in Guam, but efficient air conditioners are very expensive, and the demand for rebates for these and other technologies is always greater than the supply. There is also a risk that companies that honor the rebates will raise their prices during the duration of the rebate program.

Rebate programs can be useful but have limited impact and are not sustainable. Another alternative that can be developed in parallel with rebate programs is to establish a revolving loan fund (RLF) that can continue in perpetuity because it is financially self-sustaining.

Demand-side management (DSM) revolving loan programs allow electricity customers to borrow money to purchase energy-efficient appliances or pay for other energy-saving measures, such as weatherization improvements or cool roofs. The fund is repaid over time, typically out of the money saved by those customers on their utility bills. This allows the fund to make additional loans to other customers.

Fund Structure

There are several ways to structure an RLF, including the following.

- The utility company can lend the money and require customers to keep paying the same monthly bill or add a small premium to the bill, even though their consumption of electricity is lower, until the loan is paid off.

- An independent entity can lend the funds, with the utility company repaying the lending entity through savings on a customer's utility bill.
- A financial institution can establish a line of credit specifically for energy-saving measures, with customers being required to make regular payments until their loan is paid off. This option has two disadvantages: (1) customers with low credit scores tend to be excluded unless another entity offers to be a loan guarantor, and (2) these funds tend to be more expensive to administer due to the higher cost of recovering unpaid debts.

Financial Considerations

There are a variety of other financial considerations involved with setting up an RLF, such as ensuring that the program maintains a minimum level of funds in the bank in order to accrue interest, which can then be used to cover fund administration expenses and bad debts.

It is a good idea to set a dollar limit for the loan based on an individual's creditworthiness (FICO score) and customer payment history (from GPA records). Setting an overall dollar limit per individual will prevent people who own multiple properties, such as apartment complexes, from cornering the market.

Eligible Technologies

There are also technology considerations with setting up a DSM RLF. It is a good idea to limit eligible technologies, requiring customers to choose from a preapproved list of technologies. Guam's DSM RLF could be implemented in phases, with the first phase providing loans only for those technologies that are relatively inexpensive and provide a quick return on investment, such as cool roofs, weatherization improvements, and two-pane windows.

The second phase could include more expensive equipment and appliance purchases, such as solar hot water systems and efficient air conditioners and refrigerators. Installations of larger equipment such as roof-mounted solar hot water systems also incur the expense of ensuring that the installation meets safety standards and building codes.

Structuring the RLF in this way has the benefit of ensuring that more people can tap into the fund at its inception.

Educating Customers

A well-designed DSM RLF also includes an education component. As an example, customers with cool roofs need to understand the importance of regularly sweeping their roofs to prevent them from darkening and thereby reducing the energy savings resulting from the investment. Periodic inspections by a third party can help to ensure that maintenance requirements are met.

Many customers would benefit from having energy conservation "best practices" explained to them. This aspect of the DSM RLF could be combined with the energy awareness outreach program described under strategy 1.

Other Considerations

There are other issues to consider when developing an RLF, such as whether to structure the program so that renters can take advantage of it in addition to homeowners.

This could include requiring additional guarantees or collateral in order to mitigate the higher risk of lending to renters, as well as limiting their technology choices. Efficient appliances can move with the renter, whereas cool roofs can't.

It can be a good idea to include information on the electric utility bill that helps customers to see the impact of their energy-saving purchase on their monthly bill, such as including the amount of electricity consumed in the same month of the previous year.

This makes it easier for people to see how much their power bill has dropped. In Guam, water bills currently include information on past consumption but electricity bills do not.

How to Implement the Strategy

Responsible for Coordinating Implementation: GEO (Peter Calvo) and GPA (Joaquin Flores)

GEO is the primary lead for technology options.

GPA will do the data mining on a potential recipient's ability to pay (as it has existing customer delinquency records) and will be the enforcer of last resort, as it can disconnect customers who fail to meet their repayment obligations.

Both organizations will work in tandem to investigate sources of funding for implementing this strategy.

Potential Challenges

- Some revolving loan funds have failed. The fund's sustainability is essential. It is important to apply best practices when designing the program.
- Securing financing for technical advisors and to seed the revolving loan fund itself.
- Ensuring that inefficient equipment is disposed of. When someone purchases a new efficient air conditioner or refrigerator, they have a tendency to want to keep the old unit, which can increase their overall energy use. This can be overcome by making sure the old unit is decommissioned, for example by requiring that it is picked up by an authorized collection company or properly recycled in compliance with EPA regulations for scrap metal (with proof of disposal required).

Specific Actions and Timeline

1. Identify a source of expert advice — A third party needs to provide assistance on how to structure the loan fund.
2. Identify a source of funding for the advisors — OIA is not an option but DOE may be able to help.
3. Identify a source of funding for the loan program — Approach DOE as it is supportive of RLFs.
4. Timeline — An RLF takes about a year to set up after funding is secured. After designing the program, it needs to be tested before it is rolled out.
5. Design an energy auditing program — Including the standards that will be used, the level of audit (probably just a level 1 "walk-through" audit), and the development of an education component. The auditing program can be designed while the revolving loan fund is being developed.
   - The auditor can recommend "best practices" for people with poor energy-use habits and explain the bottom-line savings for changed behavior (e.g., "this is how much you will save if you turn off the air conditioner during the two hours it takes you to go to the store").

- It is important to capture existing behavior so that it is possible to measure the realized savings, which will encourage buy-in from the homeowners.

6. There are currently very few energy auditors in Guam. CIS has a program to train auditors but it is limited in scope. If NREL were engaged, it could help develop a "train the trainers" program tailored to Guam and the specifics of its housing stock.

Reporting on Progress and Measuring Impact

GEO and GPA will report progress at the general GETF meeting scheduled for June 14, 2013.

The impact of the program will be measured in terms of reduced energy use by participants in the RLF.

Strategy 3. Waste-to-Energy Options

Waste-to-energy (WTE) power generation could be useful on Guam because it provides baseload power and could make productive use of some of the trash that is currently disposed of in landfills. However, WTE development is a longer-term strategy because of existing legislative hurdles and the need to conduct waste characterization studies before the most appropriate WTE technology is chosen.

The GETF also supports waste reduction as a long-term strategy for Guam. Encouraging people to produce less waste conflicts with potential WTE power plant feedstock requirements, so these conflicting goals must be evaluated together.

Before exploring WTE options in more detail, the GETF decided that it is important to take the following actions.

A. Investigate Guam's Existing Waste-to-Energy Legislation

Public Law (PL) 25-175, which was adopted in December 2000, appears to restrict the development of WTE facilities in Guam.

For example, on page 3 of the document, it states: "No officer or agency of the government of Guam shall implement or expend funds, or commit resources to implement any portion of the Integrated Solid Waste Management Plan for the

Island of Guam which is disapproved or deleted by this Act or any waste-to-energy facility, or any incineration project aimed at reduction of municipal solid waste."

The full text of the law is at: http://guamlegislature.com/PublicLaws 25th/P.L.%2025-175.pdf.

The GETF will investigate the original intent of the law, its current interpretation and the likelihood of clarifying or amending it to permit WTE facilities in Guam. It is believed that many legislators would like to see WTE projects on the island, so there may be willingness to overcome this apparent obstacle to development.

GETF members will have meetings with legislators (including the legislators who originally sponsored PL 25-175), the EPA (to ascertain their current working definition of WTE), and the federal judge who would adjudicate any challenge to PL 25-175. Other legislation that may impact WTE development will also be investigated.

Responsible for Coordinating Implementation: Senator Ada's Office (Cyrus Luhr)

Potential Challenges

No challenges are anticipated as this strategy is investigative in nature, requiring an investment of time but no direct financial outlay.

Specific Actions and Timeline

1. Ask EPA's lawyers how they define waste-to-energy in general and incineration in particular.
2. Ask some of the senators who authored/supported PL 25-175 what their intent was.
3. Summarize the potential legislative barriers to WTE arising from these conversations (as part of task 3.B.2 below).
4. Formulate a legislative action plan to overcome those barriers.

Reporting on Progress and Measuring Impact

Cyrus Luhr will report progress at the general GETF meeting scheduled for June 14, 2013.

As this strategy is purely investigative in nature, it will not have any impact on fossil energy consumption. It could, however, lead to energy projects depending on the outcome of the investigation.

B. Review the Status of Waste-to-Energy Power Generation Technologies

WTE power generation encompasses a wide variety of technologies, including anaerobic digestion of organic wastes and sewage, landfill gas (methane) capture, direct combustion (incineration), cofiring with other fuels (such as biomass or coal), gasification (to produce a combustible gas), and pyrolysis (producing bio-oils that can be used to generate electricity). Not all of these technologies are equally viable or appropriate for Guam.

The GETF will investigate the various WTE technologies with possible assistance from NREL.

Responsible for coordinating implementation:
Cyrus Luhr

Potential Challenges

OIA does not have funds to support this task as it has already committed its technical assistance (TA) funds for Guam to other projects. However, this is not a significant obstacle, as this strategy is about investigating options, which can be done at very little cost.

Specific Actions and Timeline

1. Obtain information on WTE technology options and summarize this information in a report to the GETF. NREL will help with this activity.
2. Prepare a briefing document for legislators: (1) describing existing legal barriers to the adoption of WTE in Guam with background information on how these barriers came about (see task 1.A.3 above), and (2) a summary of those WTE technologies that are most applicable in Guam.

Reporting on Progress and Measuring Impact

Cyrus Luhr will report progress at the general GETF meeting scheduled for June 14, 2013.

As this strategy is purely investigative in nature, it will not have any impact on fossil energy consumption. It could, however, lead to energy projects depending on the outcome of the investigation.

C. Evaluate Other Waste Initiatives

The GETF will gather information on other waste-related initiatives and issues, including the following.

- The University of Guam's Center for Island Sustainability (CIS) is preparing to conduct a study that will examine what happens to waste in Guam. This waste stream study could help inform future decisions about the optimal WTE technologies to pursue.
- Possible synergies with, and impacts of, existing recycling/no littering campaigns, such as the i*recycle program.
- The potential impact of changes in the way the military disposes of its trash. Currently, the military is required to landfill all of its trash but that requirement will end in a few years.

Responsible for Coordinating Implementation: CIS and Edwin Reyes with Assistance from Peggy Denney

Potential challenges

No challenges are anticipated as this strategy does not require any direct financial outlays.

Specific Actions and Timeline

1. Gather information on existing and planned waste/recycling initiatives in Guam and summarize this information in a report to the GETF.
2. If the legislative hurdles can be overcome, conduct an economic analysis of the leading technology options for Guam, including relative costs and benefits. NREL may be able to provide input into this analysis.

Reporting On Progress And Measuring Impact

CIS will report progress at the general GETF meeting scheduled for June 14, 2013.

As this strategy is purely investigative in nature, it will not have any impact on fossil energy consumption. It could, however, lead to energy projects depending on the outcome of the investigation.

STRATEGY 4. TRANSPORTATION OPTIONS

The GETF considered several changes that could cut fossil fuel consumption in the transportation sector, including various approaches to reducing miles traveled, improving the fuel economy of vehicles, encouraging the use of flex-fuel vehicles by using vehicle registration rebates, and introducing electric vehicles to Guam. After careful consideration, the task force decided to focus on the following near-term tasks. Various other options will also be investigated by individual task force members, such as creating bus hubs with associated "park and ride" car parks.

A. Implement Anti-Idling Legislation to Stop Heavy Vehicles from Wasting Fuel

Tour buses idle for hours to maintain cool conditions inside the vehicle, as the engine is required to provide power to the air conditioners. Two legislative solutions were discussed by the GETF.

1. Require tour companies to install auxiliary power units (APUs) on their buses as is being done by long-distance trucking companies on the mainland to prevent truckers from running their engines for hours when sleeping just to keep the air conditioners going.
   - In the short term, APUs are more expensive than using extra fuel, but the EPA or another agency may be willing to subsidize part of the cost of the units.
   - It may be possible to make a financial argument to the tour companies along with implementing the legislation. Investigate the bottom-line cost to a company of idling: what does an hour of idling time translate into in terms of miles traveled? For some vehicles, an hour of idling is equivalent to 50-60 miles of wear on the engine, which increases maintenance costs.
2. Implement a tax targeting tour bus companies and delivery trucks as a disincentive to idle.

- Possibly add a fee to tour company registration and define a threshold for idling. If a company can prove it has reduced its idle time below the threshold, it would get the fee back.

- A fuel tax targeting diesel would not be fair, as this wouldn't just affect tour buses and delivery vehicles.
- Some vehicles (e.g., with hydraulic lifts) may need to idle to operate their equipment. Delivery companies should be given the opportunity to prove why they need to idle and would then get an exemption sticker.

Police officers could ensure compliance, or tour companies could install GPS devices that record engine idle time, as is already done on GPA vehicles.

Responsible for Coordinating Implementation: Joe Santos and Cyrus Luhr

Potential Challenges

This strategy is initially investigative in nature. The likelihood of encountering obstacles will be determined at the point that draft legislation is shown to legislators. If they are very resistant, it could be helpful to educate them on the potential advantages.

Specific Actions and Timeline

1. Establish the number of tour buses and other heavy vehicles on Guam to determine the extent of the problem.
2. Talk to tour companies to find out if they have calculated the cost of idling a bus. What does an hour of idling time translate into in terms of miles traveled or specific maintenance costs. *[Joe Santos]*
3. Draw up sample legislation and show it to legislators to see what the response is likely to be. *[Cyrus Luhr]*

Reporting on Progress and Measuring Impact

Joe Santos and Cyrus Luhr will report progress at the general GETF meeting scheduled for June 14, 2013.

B. Investigate the Pricing Structure for Vehicle Registration Fees

Fuel economy and emissions are related. If an engine is not tuned correctly, uncombusted fuel goes out the tailpipe as particulate emissions.

There are currently no emissions tests on Guam.

Vehicle registration fees are currently very low. The GETF will investigate the possibility of:

- Raising the fees significantly but provide a discount on registration fees for low-emissions vehicles or vehicles that get high miles per gallon. (The definition of "low emissions" and "low fuel consumption" will be clarified as part of the task.)
- Imposing a one-time fee at the time of purchase for vehicles with high fuel consumption.
- Using energy savings to encourage additional energy savings, e.g., mandating that all fees collected go into a revolving energy fund that the government can't tap into for other purposes.

In Guam, power plant emissions fees are paid based on the type of pollutant and the amount of each pollutant emitted. It might be possible to mirror this fee structure with prospective vehicle emissions standards.

Investigate the pricing structure for vehicle registration fees to see if (1) the base registration fees should be increased, (2) new fees should be added to change behavior.

Responsible for Coordinating Implementation: Joe Santos And Cyrus Luhr

Potential Challenges

This strategy is initially investigative in nature. The likelihood of encountering obstacles will be determined at the point that draft legislation is shown to legislators. If they are very resistant, it could be helpful to educate them on the potential advantages.

Specific Actions and Timeline

1. Investigate the current pricing structure for vehicle registration fees in other states and do a literature search on the impact of registration fees on buying habits.
2. Estimate the likely impact in Guam of (1) increasing the base registration fees, and (2) adding other fees.
3. Draw up example legislation and show it to legislators to see what the response is likely to be. *[Cyrus Luhr]*

Reporting on Progress and Measuring Impact

Joe Santos and Cyrus Luhr will report progress at the general GETF meeting scheduled for June 14, 2013.

C. Examine the Feasibility of Bringing Electric Vehicles to Guam

Electric vehicles (EVs) use much less fossil fuel than traditional vehicles, getting in excess of 100 miles per gallon equivalent (MPGe). If electricity is generated from fossil-fuel sources, EVs still use some fossil fuel, but switching the vehicle fleet to EVs could have a large impact on Guam's fossil-fuel use in the long term.

EVs have the advantage of having an onboard energy storage system (their batteries) that in some cases can be used to help stabilize intermittent power production/demand, during periods when the vehicles are plugged into the grid. Adopting EVs could thus serve two purposes.

While widespread importation of EVs is clearly a long-term goal, it might be possible to develop a pilot project through GPA or DPW to test the feasibility of using EVs in Guam. Guam has very consistent temperatures year-round, and has shorter driving distances and shorter driving times than the mainland, all of which significantly extend the overall life of EV batteries.

Responsible for Coordinating Implementation: Allison Rutter

Potential Challenges

This strategy is entirely investigative in nature and is not likely to face any obstacles.

Specific Actions and Timeline

1. Conduct a literature search of existing EV studies. *[Allison Rutter]*
2. Evaluate how the results of these studies need to be modified/tailored for Guam. *[Allison Rutter and James Thomas]*
3. Investigate existing tax incentives for purchasing EVs. *[Allison Rutter and James Thomas]*
4. Identify funding for an RFP to develop a Guam-specific study on (1) EV fleets, and (2) using EVs for grid buffering, to establish the likely impact on fossil-fuel consumption in Guan. *[Joaqin Flores and John Cruz will lead, James Thomas will provide technical input]*

Reporting on Progress and Measuring Impact

Allison Rutter will report progress at the general GETF meeting scheduled for June 14, 2013.

APPENDICES

A.1. Guam Energy Task Force Members and Other Contributors to the Action Plan

Peter Calvo
Co-Chair, Guam Energy Task Force
Director, Guam Energy Office

Reviews energy consumption on Guam, working to maximize energy conservation and efficiency, promote renewables and alternative sources of energy locally. Expedites and directs required activities to encourage energy awareness and programs.

Robert A. Underwood
Co-Chair, Guam Energy Task Force
Tom Ada
Senator, 32nd Guam Legislature

Responsible for formulating, introducing, and advocating the passage of legislation. Chairman of the Committee on Public Safety, Infrastructure, and Marine Transportation

Heidi Ballendorf
Public Affairs Director, Consolidated Commission on Utilities, Guam Power Authority

Responsible for all external messaging for both the GPA and GWA as well as the Governing Body, the CCU.

Anthony Barcinas Rudy Bautista Laura Biggs
Assistant Professor of Extension and Education Services, University of Guam Sea Grant Program

Focuses on Sea Grant program development in addition to education and extension program design and implementation. Education and extension programs serve the needs of the community in four main areas: hazard resilience, healthy coastal ecosystems, sustainable development and safe, sustainable seafood supply.

Dan Cook

Commander, United States Navy, Civil Engineer Corps

Serves as the Assistant Regional Engineer (J44) in Joint Region Marianas, Guam, assigned to NAVFAC Marianas, providing leadership and programmatic direction in the execution of over $280 million of installation support funding to Naval Base Guam and Andersen Air Force Base. Commander Cook manages the region's energy program and leads infrastructure investment efforts.

John J Cruz, Jr.

Manager, Strategic Planning & Operations Research Division (SPORD),Guam Power Authority

Oversees GPA's smart grid program, renewable energy acquisition, large customer demand-side management programs, fuel planning and contracting, outsourcing , strategic and business planning, integrated resource and other long-range planning.

Peggy Denney

*Program Administrator, i*recycle*

Promotes the recycling of aluminum beverage containers for the financial benefit of Guam's schools, working for seven local companies that created this nonprofit program. Promotes zero waste at major events and works with government agencies, businesses, organizations and others in order to increase the amount of waste being diverted from the Layon landfill.

Joseph M. Diego

Area Director, USDA Rural Development

Oversees Agency mission responsibilities for Guam, the CNMI, Republic of Palau and the Republic of the Marshall Islands, reporting to the State Director located at the State Office in Hilo, Hawaii.

Joaquin C. Flores

General Manager, Guam Power Authority

In 2005, Mr. Flores was appointed General Manager of the Guam Power Authority. Since then, he has been instrumental in charting the Authority's new direction forward to acquire renewable energy technologies through an integrated resource plan, managed the Authority's effort to regain investment grade and the inclusion of SMART Grid technology within the Authority's power grid. Mr. Flores recently presented to the Consolidated Commission on Utilities a recommendation to identify liquefied natural gas as an alternative energy resource in place of existing fuel-oil-fired steam engines.

Bruce Fredrick
General Manager, Utilities Services Specialists, Inc.

William H. Hagen
Owner, Pacific Solar & Photovoltaics, Inc.

John Heckmann Lucy Kono
Administrative Officer, Guam Energy Office
Coordinates and administers all federal grants that are awarded to GEO. Ensures compliance

Cyrus Luhr
Senior Policy Analyst, Office of Senator Thomas C. Ada
Develops policies and legislation. Manages research and reports for the Committee on Public Safety, Infrastructure & Maritime Transportation.

Michael Makio
Principal Architect and LEED AP, Taniguchi Ruth Makio Architects
A University of Oregon graduate and one of the first island professionals to achieve LEED Accredited Professional (AP) status, Michael is the principal architect for the first LEED Gold certification on Guam (GCC Learning Resource Center) and the first commercial LEED Gold certification (for the Coast 360 Federal Credit Union). Michael currently has three design projects undergoing LEED Certification and designed energy efficiency upgrades for 52 GovGuam facilities as part of the ARRA program.

Desiree Masterson
Regional Energy Program Manager, US Navy, Joint Region Marianas
The Energy Program Manager for all Navy and Air Force facilities on Guam. Coordinates with the installations to develop plans and strategies to meet Federal, DoD and Navy energy and water goals and mandates. Serves as the Regional Subject Matter Expert for facilities energy issues.

Dr. Unaisi Nabobo-Baba
Professor of Education, University of Guam, Chair Admin/Supervision

Artemio S Perez John Peterson
Assistant Vice President, University of Guam Graduate Studies and Office of Sponsored Research Programs

Edwin Reyes
Sustainability Coordinator, University of Guam, Center for Island Sustainability
Point of Contact for the Center for Island Sustainability, UOG Green Initiative Coordinator and Coordinator of Education and Outreach.

Allison Rutter
Principal, Guam Sustainability Solutions, LLC
Allison Rutter helps architects, building owners, contractors and communities create highly efficient buildings that save both energy and money. As a former member of Rocky Mountain Institute's LEED Certification Team, Allison reviewed and approved LEED projects for certification by the US Green Building Council. Allison holds a BS in Mechanical Engineering and is a licensed Professional Engineer in the state of Colorado.

Joseph Santos Gabriel Simon
Director of Engineering Development, P. H. R. Ken Micronesia, Inc.
Engineering oversight for the Ken Corp holdings on Guam: the Sheraton, Hilton, Pacific Islands Club, Hyatt and the Nikko Hotels. Focuses on optimization of engineering systems operation and energy efficiency upgrades. Involved in capital improvements, new construction and renovations. Liaison with government regulatory agencies, and engineering and technical representative for Ken Corp.

Carl Swanson James R. Thomas Elvie Tyler
Sustainability Coordinator, University of Guam, Center for Island Sustainability
Point of Contact for the Center for Island Sustainability, UOG Green Initiative Coordinator and Coordinator of Education and Outreach.

Aline Yamashita Brent Wiese
Chairman, Guam Building Code Council (GBCC)

GBCC was charged by the Legislature to create an energy code for Guam. The Guam Tropical Energy Code should soon be approved for use on island. Brent has over 30 years of architectural experience and is currently an Architect and Principal with RIM Architects (Guam), LLC. Brent is LEED AP BD+C certified.

A.2. Fossil Fuel Consumption Scenarios: Wedge Analysis

The purpose of this "wedge" analysis is to show, in graphical form, the likely impact of measures taken to reduce the use of fossil fuels. After an alternative technology or policy is implemented, it creates an irregular-looking wedge on the graph, hence the name given to this type of analysis.

General Methodology and Assumptions

The wedge analysis for Guam was performed by NREL using information provided by the Guam Power Authority and other official sources. The first wedge analysis was developed for Guam's strategic energy plan using 2010 data. That analysis established a business as usual (BAU) or base case against which the impact of any actual and proposed changes to Guam's energy consumption patterns can be compared. Those changes could be the result of energy efficiency measures, energy conservation, or expanded use of alternative power-generation technologies.

Energy can be measured using a variety of units, such as kilowatt-hours or kWh (for electricity) and British thermal units or Btu's (for heat energy). Units of energy can readily be converted into each other.

As this action plan focuses on fossil-fuel reduction, the analysis illustrates the consumption of electricity in barrels of oil equivalent (BOE) through 2020, a unit of measure that indicates how many barrels of oil would need to be consumed to meet the demand for energy. The number of Btu's in a barrel of oil is calculated for each year from data supplied by GPA and varies depending on the particular mix of fuel oil consumed by GPA that year.

Base Case

In the base case (figure 3), it is assumed that no significant steps are taken to curb fossil fuel use. In this scenario, energy consumption is projected to grow at a linear rate of 0.119% per year from 2010 onwards. That growth rate is calculated from utility data on actual consumption patterns from 2000 to

2010. This BAU scenario also assumes no major fluctuations in fuel prices by incorporating the average fuel cost for the period from 2000 to 2010.

The base case graph is a solid blue color, reflecting the fact that Guam's electricity is generated mainly from fossil fuels.

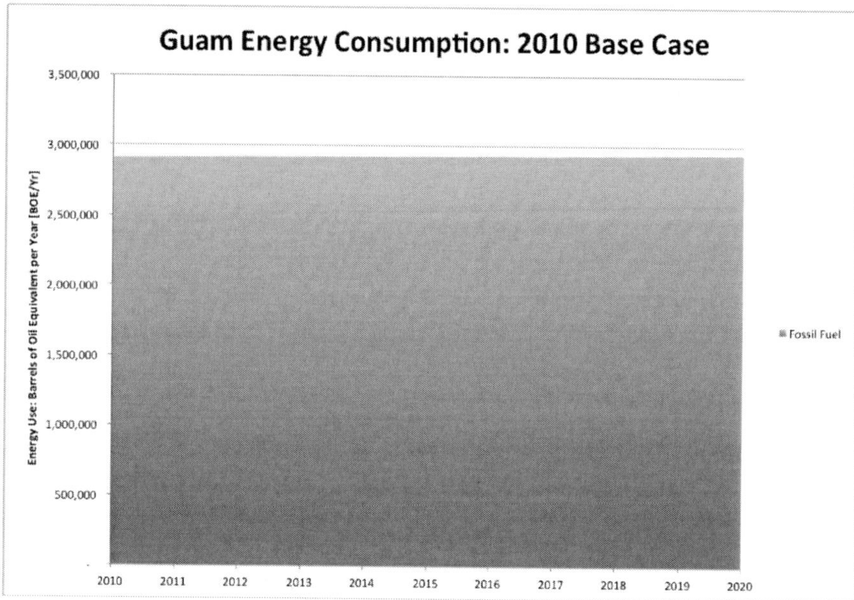

Figure 3. Energy consumed to generate electricity: 2010 base case.

Future Scenario

In the future scenario (figure 4), the overall shape of the graph reflects the same growth in total energy consumption (0.119% per year) as the base case. The graph has been updated with actual energy consumption and generation data from 2010 to 2012. The purple area at the top of the graph shows the drop in energy demand during those two years.

Projected data is used for the period from 2012 to 2020. Future projections can derive from communication and published information from GPA and other official sources regarding planned conversion of electricity generators to LNG, legally binding contracts or laws governing additions of renewable energy generation to the electricity grid, the anticipated impact on energy demand of energy efficiency and conservation measures stemming from this action plan, and the latest information on the likely impact of military personnel on Guam's energy consumption.

The actual criteria included in the future projection are described in the methodology and assumptions section below the graph. The actual energy demand in 2012 is projected forward to 2020 as a fixed percentage of the total BAU energy demand, which means that the thickness of the "demand reduction" wedge is a fixed proportion of BAU demand.

Both actual and projected data can be updated each year as more information becomes available.

The future scenario graph shows an expanding wedge of renewable energy generation replacing fossil fuel generation starting shortly after 2012. This wedge is primarily the result of Public Law 29-62, which requires that 5% of net electricity sales come from renewable sources by the end of 2015. The scenario does not currently include any energy efficiency actions.

Because of inefficiencies in the process of power generation and transmission as well as other system losses, approximately 1.136 MWh of electricity must be produced for every 1 MWh of electricity that is sold.

Replacing 5% of electricity sales from fossil-fueled generators with electricity from renewable resources would lead to a 136,146 bbl[2] reduction in fossil fuel consumption in Guam, saving approximately $15 million per year in fuel costs.

Methodology and Assumptions for the Future Scenario

This section walks through the key calculations used to develop the graph in figure 4.

Assumptions Included in the Future Scenario

- According to Public Law 29-62, 5% of net electricity sales must come from renewable sources by December 31, 2015[3]
- These sales are assumed to derive from the following:
- 429 kWh from net-metered wind and PV systems[4]
- 9.35 MW of wind generating capacity from Pacific Green Resources LLC projects
- Additional unspecified projects required to reach the goal of 5% of net electricity sales generated by 2015.

Included in the 2013 GPA integrated resource plan, but not included in wedge analysis due to the absence of an accepted RFP or a firm implementation plan (may be included in the wedge analysis at a future date):

- 10 MW geothermal plant planned for 2019
- 60 MW - 120 MW combined-cycle gas power plant

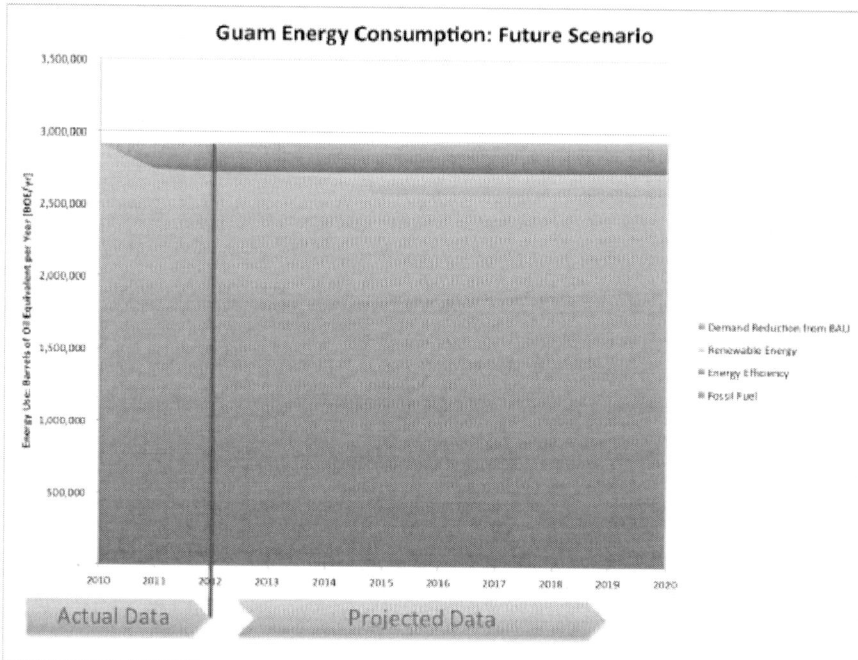

Figure 4. Energy consumed to generate electricity: actual and projected.

Calculation of Fuel Savings from Renewable Energy

Calculations begin with the assumption that 5% of net electricity sales come from renewable energy by 2015 and use 2012 as the baseline year.

The following data was provided by the utility:

2012 Fuel Consumption Data			
Fuel	Units bbl5)	Costs	Million Btu (MMBtu)
High Sulfur RFO	1,928,688	$212,876,225	12,150,734
Low Sulfur RFO	734,015	$78,093,042	4,624,295
Ultra Low Sulfur Diesel	60,208	$9,095,427	351,494
TOTALS	2,722,911	300,064,694	17,126,523

2012 Fuel Cost Data	
Fuel Type	Cost I$/bbl]
High Sulfur RFO	110.37
Low Sulfur RFO	106.39
Ultra Low Sulfur Diesel	151.07

2012 Fuel Generation Data	
Total Electricity Generation [MWh]	1,784,446
Total Net Electric Sales [MWh]	1,570,396
Generation Efficiency (Heat Rate) [MMBtu/MWh]	10.145

In 2012, GPA generated 1,784,446 MWh of electricity, of which 1,570,396 MWh was sold.

The ratio of generated power to net sales can be used to derive a single number that summarizes the approximate transmission and distribution system losses, losses in the electricity generating process and other losses associated with the avoided energy generation:

1,784,446 generated MWh/1,570,396 sold MWh
= 1.1363 generated MWh/sold MWh

In other words, 1.1363 MWh is generated for every 1 MWh sold.

To calculate 5% of net electricity sales:

1,570,396 MWh * 0.05 = 78,520 MWh

To calculate the fuel savings from a 5% reduction in electricity consumption:

It seems reasonable to assume a linear relationship between electricity consumed and electricity generated. It also seems reasonable to assume a linear relationship between electricity generated and the fuel used to generate that electricity. The model is evaluating electricity generation (and fuel savings) at the margin, and there is no reason to assume that power plant efficiency will be impacted significantly by an incremental reduction in power plant output.

Thus a 5% reduction in electricity sales results in a 5% reduction in electricity generation and a 5% reduction in fossil fuel consumption for power generation.

Total fuel consumed by GPA for electricity production in 2012 was 2,722,911 bbl. The projected fuel savings from a 5% reduction in electricity consumption is thus:

2,722,911 bbl * 0.05 = 136,146 bbl

To calculate the dollar value of the fuel savings:

The weighted average cost of fuel is calculated using the following information provided by GPA. The weighting of each fuel is calculated as the percentage of individual fuel type relative to the total barrels of oil purchased.

High Sulfur RFO: 70.83%
Low Sulfur RFO: 26.96%
Ultra Low Sulfur Diesel: 2.21%

The weighted average cost (C_w) is then calculated as:

C_w = (\$110.37*0.7083) + (\$106.39*0.2696) + (\$151.07*0.0221)
= \$110.20/bbl

The fuel cost savings (rounded to the nearest dollar) is calculated using the projected fuel savings and the weighted average cost of fuel for 2012:

136,146 bbl * \$110.20/bbl = \$15,003,289

A.3. Glossary of Energy Management Concepts

Demand-side management (DSM) — Actions undertaken by a utility to change the level or timing of energy use on the customer side of the electricity meter, generally with the intention of optimizing existing and planned generation and transmission assets on the utility side of the meter. It encompasses utility-initiated actions to improve energy efficiency, increase energy conservation, and reduce peak electricity demand. Examples include rebates, incentives and utility investments that improve building shell insulation and the efficiency of heating and cooling systems, and rate structures that shift demand from times of peak energy use to off-peak hours.

DSM includes other load management approaches, such as direct load control systems that enable the grid operator to regulate the electricity used by individual appliances or equipment on customer premises (usually residences), and contractual arrangements that allow the utility to reduce demand by specific customers (usually businesses) through direct action by the grid operator (remote tripping) or by the customer voluntarily taking action to reduce demand when requested by the grid operator.

DSM is sometimes defined to include utility incentives to encourage electricity generation by customers, because it reduces the need for the utility to invest in power plants or transmission and distribution networks. This definition of DSM includes solar photovoltaic (PV) or other distributed generation systems installed on a customer's property, but does not include PV or other distributed generators installed on the transmission or distribution grid.

DSM refers only to energy and load-shape modifying activities that are undertaken in response to utility-administered programs, such as a utility incentive to customers encouraging them to buy more energy-efficient light bulbs. It does not refer to changes arising from normal operation of the market (such as the independent decision by a customer to buy such light bulbs because electricity is expensive) or from government-mandated energy efficiency standards.

Energy conservation — The process of using less energy by reducing or going without a service. Energy conservation typically requires a change in behavior or not using an existing technology, e.g., turning off a light, using the stairs instead of taking the elevator, or turning up the thermostat on an air conditioner. Popular approaches to accomplishing energy conservation goals include: (1) educating people about how to conserve energy, and (2) implementing nonenergy-consuming technologies such as cool roofs, which reduce the need for air conditioning.

Energy efficiency — The process of using less energy to provide the same service, or using the same amount of energy to provide more services. Energy efficiency typically requires the use of a different energy technology or using existing technology in a different way. For example, a compact fluorescent light (CFL) or light-emitting diode (LED) bulb is more energy efficient than a traditional tungsten incandescent light bulb because it produces the same amount of light (measured in lumens) using significantly less electrical energy.

Supply-side management — Refers to the practice of electric utilities building generating plants to serve whatever demand customers require. This practice was the norm until the 1970s.

End Notes

[1] Ian Baring-Gould et al., Guam Initial Technical Assessment Report. National Renewable Energy Laboratory, NREL/TP-7A40-50580, April 2011.

[2] barrels

[3] http://www.guampdn.com/article/20130423/NEWS01/304230310/Wind-turbine-added-Uni ver sity-Guam, accessed April 24, 2013

[4] Email to NREL from GPA.

[5] barrels.

INDEX

A

abatement, 99
access, 42, 47, 49, 52, 89, 90, 98, 118
accounting, 6, 119
ADA, 134
adhesion, 81
administrators, 40
adults, 39
advocacy, 44, 110, 111
age, 19
agencies, 23, 35, 91, 112, 118, 137, 140, 141, 158
agricultural sector, 60
agriculture, 41
air emissions, 62
Air Force, 49, 58, 71, 158, 159
algae, 58, 81
alternative energy, 66, 108, 159
American Recovery and Reinvestment Act, 26
American Recovery and Reinvestment Act of 2009, 26
American Samoa, vii, 1, 109, 139
anaerobic digesters, 59
anaerobic digestion, 63, 151
aquaculture, 41
architect(s), 65, 85, 118, 159, 160
ARs, 15
Asia, 55, 62
assessment, viii, 2, 3, 6, 8, 24, 34, 40, 41, 45, 46, 47, 53, 61, 62, 63, 65, 99, 100, 109, 114, 119, 127, 138, 140
assets, 128, 166
atmosphere, 66, 108
attachment, 50
audit(s), 34, 35, 36, 38, 39, 65, 76, 78, 148
authority, 141
automation, 70
autonomy, 110
awareness, 35, 36, 39, 41, 65, 77, 78, 81, 84, 86, 110, 111, 118, 141, 147, 157

B

background information, 6, 50, 151
banks, 82, 117
barriers, 3, 4, 8, 41, 107, 116, 120, 150, 151
base, 14, 19, 58, 63, 67, 69, 70, 71, 73, 74, 129, 132, 155, 161, 162
batteries, 156
BD, 161
benchmarking, 25, 37
benefits, 25, 33, 54, 61, 90, 116, 118, 141, 152
biodiesel, 54, 57, 58, 62, 63, 107, 133
bioenergy, 60
biofuel, 54, 58, 62, 64
biogas, 54, 60
biomass, 8, 25, 54, 60, 61, 63, 64, 65, 151

blends, 57, 63
BOE, 129, 161
boilers, 54, 57, 58
breakdown, 10, 13, 23, 31, 32
building code, 3, 27, 35, 37, 46, 75, 77, 85, 86, 146
burn, 55
business model, 44
businesses, 72, 76, 117, 141, 158, 167
buyer, 128

C

campaigns, 3, 141, 142, 143, 144, 152
Caribbean, 45
case studies, 62
case study, 49, 62
cash, 29
cash flow, 29
category a, 22
Census, 101, 112
certification, 78, 114, 139, 159, 160
challenges, 28, 54, 60, 61, 63, 66, 108, 116, 144, 150, 152
children, 39, 86, 143
China, 53
CIA, 4, 101, 102
circulation, 48
CIS, 142, 144, 145, 149, 152
citizens, 119, 141
classes, 47
clean energy, 8, 25, 41, 107, 114, 115, 126, 138, 139
clean technology, 57
cleaning, 36
climate(s), 11, 12, 20, 27, 36, 74, 79, 80
closure, 29, 63
CNMI, 101, 103, 109, 139, 158
coal, 8, 151
Coast Guard, 49, 50, 103
coastal ecosystems, 158
coatings, 81
collaboration, 46, 47, 141
collateral, 147
color, 162
combustion, 14, 19, 55, 57, 69, 70, 71, 151
commercial, 8, 12, 18, 20, 21, 22, 27, 33, 49, 50, 55, 58, 60, 61, 65, 75, 77, 78, 159
Commonwealth of the Northern Mariana Islands, vii, 1
communication, 44, 162
communities, 25, 28, 36, 39, 44, 54, 60, 62, 86, 112, 117, 126, 137, 160
community, 19, 39, 41, 42, 44, 47, 76, 78, 80, 100, 108, 114, 118, 126, 127, 128, 139, 157
community support, 128
compensation, 74
competition, 108
competitive process, 45
complement, 19
compliance, 29, 77, 134, 148, 154, 159
composition, 28, 31, 32, 55, 61
computer, 73
computer systems, 73
conditioning, 3, 23, 25, 36, 37, 38, 39, 48, 76, 78, 80, 84, 86, 107, 167
configuration, 90
consensus, 118, 140
consent, 29
conservation, ix, 20, 21, 35, 39, 66, 73, 76, 77, 78, 86, 89, 113, 135, 137, 138, 139, 144, 162, 167
conserving, 21
construction, 4, 10, 19, 20, 27, 29, 30, 33, 36, 37, 40, 44, 58, 63, 65, 74, 75, 78, 84, 85, 86, 88, 90, 118, 128, 129, 160
consulting, 29
consumers, 12, 18, 25, 125, 143
consumption, 3, 6, 7, 8, 10, 12, 16, 18, 20, 21, 22, 23, 24, 25, 40, 48, 72, 73, 86, 106, 112, 119, 129, 132, 145, 147, 161, 162, 165, 166
consumption patterns, 7, 119, 161
containers, 158
controversial, 34
conversations, 150
cooking, 12, 21
cooling, 21, 22, 25, 38, 40, 69, 80, 166
coordination, 133, 134

coral reefs, 8
correlation, 46, 99
cost, viii, 2, 18, 19, 22, 25, 29, 36, 38, 39, 43, 44, 46, 47, 48, 49, 54, 56, 57, 58, 60, 61, 62, 63, 71, 73, 78, 80, 84, 89, 90, 108, 114, 117, 129, 133, 138, 143, 146, 151, 153, 154, 162, 166
cost benefit analysis, 62
cost effectiveness, 89
cost saving, 166
cost-benefit analysis, 25, 38, 143
covering, 142
creative thinking, 34
creditworthiness, 146
criticism, 44
crop(s), 8, 54, 65
crop residue, 54
CT, 4, 67, 69, 70, 71
Cuba, 45
cultural influence, 12
cultural norms, 125
culture, 20, 79
cure(s), 81
current limit, 44
curriculum, 40, 77, 113, 137
customers, 14, 22, 25, 26, 73, 145, 146, 147, 148, 167

D

data collection, 6
data mining, 147
database, 77
debts, 146
decay, 59
delegates, 109, 139
delinquency, 147
Department of Agriculture, 43, 60, 98
Department of Defense, 4, 6, 119
Department of Education, 23, 25, 35, 77
Department of Energy, 4, 12, 109, 139
deposits, 73, 81
designers, 142
destruction, 20
diesel engines, 69, 70
diesel fuel, 48, 57
digestion, 54
direct action, 167
disclosure, 39
dissatisfaction, 79
distribution, 14, 19, 48, 52, 69, 70, 71, 75, 89, 90, 94, 95, 96, 97, 98, 107, 165, 167
diversification, 66, 106, 107, 110, 111, 120
diversity, 10, 25, 28, 53
DOI, vii, 1, 4, 109, 139
draft, 2, 142, 154, 155
drainage, 59
DSM, 146, 147, 166, 167
dumping, 29

E

earnings, 10
economic crisis, 73
economic development, viii, 2, 46, 113, 137, 139
economics, 7, 33, 44, 45, 46, 56, 58, 60, 62, 98, 116, 119
economies of scale, 58
education, 21, 41, 46, 77, 106, 113, 114, 137, 139, 141, 144, 147, 148, 157
educational institutions, 113, 137
educational materials, 65
EIS, 4, 10, 33, 101, 103
electricity, 10, 12, 14, 15, 18, 22, 28, 47, 54, 56, 69, 73, 84, 89, 107, 117, 129, 140, 143, 145, 147, 151, 156, 161, 162, 163, 164, 165, 166, 167
emergency, 19, 69, 70, 71
employees, 14, 73
employment, 10
end-users, 22
energy conservation, 21, 39, 78, 81, 85, 86, 88, 107, 108, 114, 116, 117, 126, 143, 147, 157, 161, 166, 167
energy consumption, vii, viii, ix, 2, 6, 11, 12, 16, 17, 18, 19, 20, 22, 23, 24, 25, 38, 39, 75, 80, 84, 86, 105, 109, 110, 119, 135, 140, 141, 143, 150, 151, 152, 157, 161, 162

energy prices, 66, 107, 108
energy recovery, 54
energy security, viii, 2, 108, 125, 126, 139, 141
energy supply, 12, 106, 107, 110, 111, 120
enforcement, 27, 35, 106, 118
engineering, 44, 128, 160
environment(s), 28, 45, 47, 66, 73, 116
environmental effects, 25
environmental impact, 43, 46, 99, 108
environmental policy, 108
Environmental Protection Agency, 4, 34, 103
environmental quality, 66
EPA, 4, 37, 59, 148, 150, 153
equipment, 3, 21, 22, 25, 26, 27, 35, 37, 39, 47, 54, 57, 62, 65, 76, 86, 90, 98, 118, 128, 146, 148, 154, 167
ethnic groups, 10
Europe, 55
evaporation, 81
evidence, 53, 58
execution, 158
Executive Order, 112, 113, 136, 137
expenditures, 113, 137, 141
expertise, viii, 2, 42
exposure, 49, 81, 141
extreme weather events, 45

F

farmers, 60
Federal Emergency Management Agency, 71
federal facilities, 113, 137
federal government, 36
federal mandate, 114, 139
feedstock, 32, 33, 54, 55, 58, 61, 63, 149
feedstocks, 54, 58, 60, 64
Filipino, 10
financial, 110, 113, 128, 129, 137, 146, 150, 152, 153, 158
financial resources, 129
flaws, 92
flex, 153
flexibility, 57, 71
fluctuations, 108, 129, 162
fluid, 47, 48
fluidized bed, 55
food, 60
food production, 60
force, 113, 139, 140, 153
fossil energy consumption, vii, viii, ix, 105, 135, 150, 151, 152
fuel cell, 25
fuel consumption, 12, 15, 38, 40, 64, 106, 108, 110, 111, 112, 120, 129, 131, 132, 140, 153, 155, 156, 163, 165
fuel efficiency, 16
fuel prices, 15, 129, 162
fuel sources, vii, viii, ix, 105, 110, 135, 156
funding, 9, 44, 76, 77, 78, 99, 113, 137, 140, 142, 144, 148, 156, 158
funds, 26, 29, 53, 146, 148, 149, 151

G

gasification, 151
General Motors, 4, 69
geography, 6, 101
GIS, 90, 91
goal setting, 3
governance, 115
government expenditure, 76
governments, 25
governor, 40
GPS, 154
grants, 77, 117, 159
graph, 129, 132, 161, 162, 163
gravity, 48
growth, 22, 64, 73, 79, 81, 112, 116, 129, 161, 162
growth rate, 161
Guam Energy Action Plan, v, vii, ix, 105, 109, 135
Guam Energy Task Force (GETF), vii, viii, 105, 136
Guam Initial Technical Assessment Report, v, vii, viii, 1, 105, 109, 129, 134, 140, 168

Guantanamo, 45
guidance, 50, 77, 112, 137
guidelines, 26, 39, 58
Gulf war, 15

H

halogen, 84
Hawaii, 16, 28, 49, 50, 57, 58, 60, 62, 103, 158
hazards, 8
health, 44, 79
higher education, 75
highways, 90
history, 146
homeowners, 147, 149
homes, 35, 42, 72, 86, 98
hospitality, 75
hotel, 35, 38, 48, 82
hotels, 12, 18, 20, 21, 35, 38, 40, 48, 49, 72, 78
housing, 50, 76, 84, 85, 88, 149
human, 54
humidity, 12, 76, 79, 80
hybrid, 25

I

ideal, 89, 90
identification, 3, 42, 47, 64
identity, 119, 141
impact assessment, 46
impact energy, 6, 10, 20, 132
improvements, 3, 20, 35, 44, 52, 59, 76, 78, 80, 86, 88, 98, 107, 129, 131, 132, 143, 145, 146, 160
income, 108
independence, 66
individuals, 109, 140
industries, 40, 48
industry, 10, 34, 36, 37, 40, 60, 64, 75, 85, 86, 106, 116
infrastructure, 10, 16, 42, 43, 44, 57, 59, 61, 73, 95, 97, 108, 113, 137, 158
inspections, 147
inspectors, 77
insulation, 36, 76, 166
integration, 28, 115, 138
integrity, 81, 116
interface, 73
investment(s), 8, 59, 61, 66, 73, 78, 106, 108, 117, 126, 143, 145, 146, 147, 150, 158, 166
investors, 25, 118
Iowa, 58
islands, 44, 54, 62, 125
isolation, 66, 82, 108, 116, 139
issues, viii, 1, 27, 29, 44, 46, 62, 73, 128, 147, 152, 159

J

Japan, 10
job creation, 46
jurisdiction, 8

L

landfills, 29, 63, 149
landscape, 25, 114, 138
languages, 10
laws, 34, 61, 162
lawyers, 150
lead, 29, 45, 60, 74, 125, 147, 150, 151, 152, 156, 163
leadership, 158
leaks, 81
LED, 5, 76, 88, 167
LEED, 5, 20, 78, 159, 160, 161
legal issues, 56
legend, 131, 132
legislation, ix, 26, 27, 28, 34, 57, 72, 135, 150, 153, 154, 155, 157, 159
lending, 146, 147
life cycle, 119
lifetime, 36, 81
light, 21, 39, 76, 82, 84, 86, 87, 143, 167
limestone, 8, 53
liquefied natural gas, 8, 107, 159
literacy, 10, 41, 113, 119, 137

lithium, 80
loans, 56, 145, 146
local community, 44
local government, 25

M

machinery, 37
magnitude, 90
major issues, 45
majority, 75
Malaysia, 58
man, 13
management, ix, 5, 12, 14, 29, 39, 73, 135, 145, 158, 166, 167
manure, 54
marketing, 142
Marshall Islands, 158
mass, 55
materials, 19, 30, 31, 33, 36, 39, 40, 61, 65, 80, 81, 86, 118, 119
matter, 61
measurement, 42, 98, 99, 112
media, 144
medical, 20, 75, 78
membership, 112, 137
memorandums of understanding, 76
messages, 39, 141, 144
metals, 31, 32
meter, 26, 139, 166
methanol, 58
methodology, 3, 6, 7, 88, 90, 128, 163
microturbines, 25
military, viii, 1, 6, 9, 10, 19, 20, 30, 33, 34, 44, 46, 59, 63, 66, 71, 73, 88, 89, 107, 108, 141, 152, 162
mission, 66, 113, 137, 158
models, 27, 37, 117
modifications, 3, 43, 57, 62, 80
modules, 53, 94
moisture, 38
mold, 37, 79, 81
momentum, 40
MSW, 5, 29, 54, 55, 56
municipal solid waste, 5, 29, 33, 150

N

National Renewable Energy Laboratory, vii, viii, 1, 5, 6, 105, 109, 134, 135, 139, 168
National Renewable Energy Laboratory (NREL), vii, 1, 6, 109, 139
natural advantages, 66, 108
natural disaster(s), 108
natural resources, 112, 137
nitric oxide, 5
NOAA, 9
non-renewable resources, 66, 108
NRCS, 5, 60

O

obstacles, 154, 155, 156
officials, 37
oil, viii, 1, 5, 14, 16, 18, 54, 57, 58, 61, 66, 70, 76, 107, 129, 134, 139, 159, 161, 166
opacity, 57
operating costs, 14
operations, 7, 12, 29, 46, 60, 69, 70, 73, 119, 129
opportunities, 2, 3, 6, 8, 20, 23, 34, 35, 42, 50, 54, 60, 61, 62, 64, 65, 76, 103, 108, 114, 116, 117, 138, 139, 144
optimization, 160
organ, 118
outreach, ix, 3, 41, 46, 65, 77, 106, 113, 114, 135, 137, 139, 141, 142, 143, 144, 147
outsourcing, 158
oversight, 160
ownership, 47, 52, 64, 91, 118
ownership structure, 118

P

Pacific, vii, 1, 8, 10, 11, 45, 75, 102, 109, 139, 144, 159, 160, 163
Pacific Islanders, 10
parallel, 145
parents, 143

participants, 31, 149
payroll, 10
permit, 150
personal benefit, 44
petroleum, 57, 63, 112, 133, 134
Petroleum, 134
photons, 47
pipeline, 134
plants, 14, 52, 54, 55, 56, 57, 58, 59, 60, 61, 63, 67, 70, 88, 167
policy, viii, 2, 8, 25, 40, 41, 64, 65, 107, 109, 113, 114, 115, 116, 117, 125, 137, 138, 139, 141, 161
politics, 126
pollution, 25, 81, 143
polyurethane, 81
poor performance, 74
population, 8, 10, 22, 30, 44, 66, 86, 112
population density, 44
population growth, 112
portfolio, 8, 20, 25, 37, 65, 132
power generation, 12, 43, 73, 110, 112, 117, 143, 149, 151, 163, 165
power plants, 16, 89, 167
precedent, 86
precipitation, 12
predators, 8
preservation, 113, 137
primary sector, 12
principles, 116
private sector, 8, 30, 110
procurement, 73, 128
production costs, 58
professionals, 159
profit, 76
project, viii, 2, 3, 6, 29, 42, 43, 44, 45, 46, 56, 59, 70, 71, 73, 80, 89, 95, 98, 99, 109, 110, 111, 117, 120, 128, 132, 134, 150, 156
propane, 12
property rights, 128
public interest, 126
public sector, 141
public-private partnerships, 117
pumps, 25, 48
pyrolysis, 151

Q

quality of life, 66, 108, 116
quotas, 116

R

radar, 43, 99
radiation, 48
radio, 119
RE, 126
real estate, 51, 76
recommendations, 6, 27, 66, 81, 88, 107, 119
recovery, viii, 2, 56, 71, 103
recreation, 41, 96
recreational, 49
recruiting, 73
recycling, 29, 30, 31, 33, 62, 141, 152, 158
regulations, 33, 49, 61, 64, 148
regulatory agencies, 160
reliability, 57, 73, 116
relief, 48
religion, 10
renewable energy technologies, vii, viii, 1, 2, 3, 7, 66, 109, 129, 132, 139, 140, 158
renewable fuel, 54
requirements, 19, 26, 27, 29, 38, 45, 50, 74, 75, 118, 147, 149
researchers, 61
Residential, 22, 24, 26, 27, 30, 36, 76, 84, 85, 91, 103
residues, 55
resilience, 157
resistance, 75
resolution, 92
resource management, 76
resources, 34, 36, 50, 54, 55, 61, 67, 77, 103, 113, 137, 139, 142, 149, 163
response, 154, 155, 167
restaurants, 58
restoration, 70, 72, 113, 137
retail, 21, 84, 101

revenue, 29, 72
risk(s), 44, 45, 46, 47, 60, 118, 126, 145, 147
risk assessment, 47, 126
routes, 134
rules, 26, 33

S

safety, 26, 133, 146
SARS, 16
savings, 27, 34, 36, 40, 46, 47, 49, 50, 72, 75, 77, 81, 86, 117, 145, 146, 147, 148, 149, 155, 165, 166
scarce resources, 108
school, 20, 35, 72, 76, 77, 86, 119, 133, 158
scope, 52, 55, 149
seafood, 158
security, 66, 73, 108, 139
seed, 148
self-assessment, 65, 118
sensors, 81, 84
September 11, 16
services, 29, 50, 71, 76, 117, 167
sewage, 59, 60, 151
shape, 16, 74, 89, 162, 167
shelf life, 58
shingles, 81
showing, 16
silica, 80
simulations, 46
sludge, 54
social acceptance, 3, 43, 46
sodium, 76, 88
software, 90
solar system, 25
solid waste, 5, 12, 29, 30, 31, 32, 33, 54, 62, 76
solution, 19, 39, 66
species, 43, 58, 64
specifications, 86
speculation, 57
spending, 10, 143
stability, 43, 108
stakeholders, vii, viii, 1, 6, 33, 105, 106, 109, 119, 126, 127, 129, 140
state(s), 25, 33, 56, 62, 66, 77, 79, 102, 149, 155, 160
statistics, 18, 50
stock, 19, 27, 37
storage, 62, 156
storms, 81
strategic planning, 47, 64, 109
structure, 28, 44, 45, 46, 81, 111, 113, 114, 120, 137, 138, 145, 147, 148, 155
sulfur, 14, 16, 57
suppliers, 36, 38
surface properties, 81
survival, 45
sustainability, viii, 2, 58, 66, 108, 113, 126, 132, 137, 139, 144, 148
sustainable development, 77, 158
sustainable energy, 112, 116, 126, 137

T

talent, 142
target, 28, 30, 39, 89, 112, 142
Task Force, vii, viii, ix, 2, 34, 35, 38, 40, 52, 57, 61, 63, 64, 65, 66, 105, 109, 112, 119, 135, 136, 141, 157
tax incentive, 156
teachers, 39
technical assistance, viii, 2, 4, 35, 41, 151
technical efficiency, 75
techniques, 43, 81
technological advancement, 84
technologies, 3, 7, 8, 36, 38, 40, 41, 46, 47, 49, 56, 61, 66, 73, 74, 81, 84, 109, 117, 126, 140, 145, 146, 151, 152, 167
technology, 3, 8, 21, 38, 40, 42, 43, 49, 53, 55, 69, 73, 74, 80, 84, 107, 108, 110, 120, 128, 140, 146, 147, 149, 151, 152, 159, 161, 167
temperature, 11, 12, 36, 53, 65, 78, 79, 80
temporary housing, 88
tensile strength, 81
tenure, 128
term plans, 58

territorial, 41
territory, 8, 66
testing, 45, 57, 133
testing program, 57
theft, 16
thermal energy, 8, 48
time frame, 107, 111, 120
tin, 76
Tony Babauta, vii, 1
total energy, 14, 24, 88, 162
tourism, 10, 16, 41, 74, 108
training, 35, 36, 40, 65, 74, 118
transesterification, 58
transformation, 25, 108, 109, 116, 125, 126, 127, 128
transmission, 14, 19, 42, 43, 44, 51, 52, 69, 70, 71, 72, 89, 90, 94, 95, 96, 97, 98, 163, 165, 166, 167
transport, 12
transportation, ix, 6, 12, 28, 65, 106, 108, 110, 111, 115, 120, 133, 135, 138, 140, 153
treatment, 54, 59, 60, 61, 63, 88
trial, 134
tropical storms, 72
tungsten, 167

U

U.S. Department of Agriculture, 5, 60
U.S. Department of the Interior, 105, 109, 135, 139
United, 8, 29, 34, 49, 55, 158
United States, 8, 29, 34, 49, 55, 158
urethane, 81
USDA, 5, 60, 86, 158

V

vegetable oil, 48, 54, 57, 58, 62
vegetation, 38, 49, 63
vegetative cover, 89
vehicles, ix, 66, 107, 110, 114, 135, 139, 153, 154, 155, 156
ventilation, 21, 38
Vice President, 160
vision, 66, 113, 126, 137
vocational training, 114, 139
volatility, viii, 1, 66, 108, 116, 139
vulnerability, 66

W

Washington, 103
waste, ix, 5, 28, 29, 30, 31, 32, 33, 34, 54, 55, 56, 58, 60, 61, 62, 103, 107, 135, 149, 150, 152, 158
waste management, 28, 29
wastewater, 54, 59, 60, 61, 63
water, 3, 8, 20, 21, 23, 25, 38, 39, 40, 41, 47, 48, 49, 50, 52, 53, 76, 80, 81, 84, 88, 107, 109, 146, 147, 159
water heater, 107
wear, 153
webpages, 142, 144
websites, 40, 106
well-being, 66
wind farm, 56
wind speed, 42, 43, 45, 98, 99
wind turbines, 42, 44, 46, 98, 99, 100
windows, 36, 38, 76, 78, 79, 82, 146
wood, 20, 54, 76
wood waste, 54
workers, 65, 88, 118

Y

yield, 55, 81